Work Study
and
Ergonomics

Work Study
and
Ergonomics

Dr. P.C. Tewari

Professor
Mechanical Engineering Department
NIT Kurukshetra, INDIA

Taylor & Francis
Taylor & Francis Group
Boca Raton   London   New York

CRC is an imprint of the Taylor & Francis Group,
an informa business

Ane Books Pvt. Ltd.

Work Study and Ergonomics

P.C. Tewari

© Author, **2018**

Published by

Ane Books Pvt. Ltd.

4821, Parwana Bhawan, 1st Floor, 24 Ansari Road, Darya Ganj,
New Delhi - 110 002, Tel.: +91(011) 23276843-44, Fax: +91(011) 23276863
e-mail: kapoor@anebooks.com, Website: www.anebooks.com

For

CRC Press

Taylor & Francis Group
6000 Broken Sound Parkway, NW, Suite 300
Boca Raton, FL 33487 U.S.A.
Tel : 561 998 2541
Fax : 561 997 7249 or 561 998 2559
Web : www.taylorandfrancis.com

For distribution in rest of the world other than the Indian sub-continent

ISBN: 978-11-3804-955-0

British Library Cataloguing in Publication Data
A catalogue record for this book is available from the British Library

Printed at Replika Press Pvt. Ltd.

Preface

This book attempts to deal with all the aspects of the Work Study and Ergonomics, in a simple, concise and lucid manner so as to develop the clear understanding of the concepts in the mind of the readers. Various topics namely Productivity, Importance of Work Study, Method Study, Work Measurement and Ergonomics have been covered with a balance of relevant fundamentals and their industrial applications. As a matter of fact, now-a-days, the industries with more complexity and automation truly demand a scientific approach like Work Study and Ergonomics to deal with the problems relating to Human Considerations, Time Standards, Wages & Incentives, Man- Machine Systems and Design of Work Place etc.

This book will be of immense value to the graduate students, studying at various Universities, IITs, NITs and other engineering colleges along with the practicing engineers and researchers working in the field of Industrial Engineering at large and Work Study and Ergonomics in particular.

Dr. P.C. Tewari

Contents

Productivity

1.1 Basic Concepts of Productivity

Productivity is defined as the efficiency of a Production System. It is measured in terms of the rate of output per unit of input. In other words it is an average measure of the efficiency of production. Efficiency in production implies a person's capabilities to generate income which is measured by the formula of real output value minus real input value. Increasing national productivity can raise people's living standards, as more real income would improve their ability to purchase more goods and services, enjoy leisure, improve housing and education standards, and contribute to social and environmental programmes. Productivity growth also helps businesses to generate more profit.

Productivity (P) = **Output / Input.**

Output implies production, while Input means land, labour, capital and management, etc. There are three perspectives that dominate the field of Productivity, namely Economics, Industrial Engineering and Administration. In fact, these perspectives have complicated a search for any precise definition of the concept 'Productivity'.

Productivity is not everything, yet in the long run it is almost everything. A country's ability to improve its standard of living over a time depends almost entirely on its ability to raise its output per worker. Despite the proliferation of computers, productivity growth was relatively slow from the 1970s through the early 1990s. Although several possible causes for the slowdown have been proposed, there has been no consensus ever. The matter has been subjected to a continuing debate that is beyond questioning: whether the computers alone can significantly increase productivity, or the potential to increase productivity has become exhausted.

Productivity is one of the main concerns of business management and engineering. Practically all companies have established procedures for collecting, analysing and reporting the necessary data regarding productivity. Typically the accounting department in all industries has the overall responsibility for collecting, organising and storing the data, though data originate in different departments.

1.2 Significance of High Industrial Productivity

Productivity is created in the real process. Its gains are distributed in the income distribution process and these two processes constitute the production process. The production process and its subprocess, the real process and income distribution process occur simultaneously, and only the production process is identifiable and measurable by the traditional accounting practices. The real process and income distribution process can be identified and measured by extra calculation, and this is why they need to be analysed separately in order to understand the logic of production performance.

Various benefits derived from higher industrial productivity are as follows:

1. It helps to cut down cost per unit, thereby increasing the profits.
2. Gains from productivity can be transferred to consumers in the form of lower-priced products, or better quality products.
3. These gains can also be shared with workers or employees by paying them higher wages.
4. An entrepreneur who gives more industrial productivity has better chances of exploiting export opportunities.
5. It generate more employment opportunities.

1.3 Factors Influencing Productivity

Following factors influence productivity:

(*a*) Human factors: Human nature and human behavior are the two most significant determinants of productivity. Human factors include:

    (*i*) Ability to work

    (*ii*) Willingness to work

(*b*) **Technological factors:** Technological factors exert significant influence on the level of productivity. These include the following:

    (*i*) Size and capacity of plant

    (*ii*) Product design and standardisation

    (*iii*) Timely supply of materials and fuels

    (*iv*) Repairs and maintenance

    (*v*) Proper production planning and command on various facts of productivity

    (*vi*) Plant layout and location

    (*vii*) Material-handling system

    (*viii*) Inspection and quality control

    (*ix*) Inventory control

(*c*) **Managerial factors:** The competence and attitudes of managers have considerable effect on productivity. Sometimes despite the use of latest and best technology, with trained man power an organisation may end up having low productivity.

(*d*) **Natural factors:** Natural factors, such as physical, geographical and climatic conditions exert considerable influence on productivity, e.g., in extreme climates productivity tends to be comparatively low. Natural resources like water, fuel and minerals also influence it.

(*e*) **Sociological factors:** Social customs, traditions and institutions influence people's attitudes towards work and their jobs. Close ties with the land and native place hamper stability and discipline among industrial labours. Strict following of social customs and traditions, prejudices against religion and caste, i.e., ethnic reasons have inhibited the growth of modern industries and technology in some countries.

(*f*) **Political factors:** Strict imposition of law and order, stability of government, good rapport among states, etc., are essential for high productivity in industries. The taxation policies of the government, demotivate capital formation, modernisation and expansion of plants and tariff policies influence competition. Moreover, elimination of sick and inefficient units also helps to improve productivity.

(*g*) **Economic factor:** Economic factor plays a major role in improving productivity. Size of market, banking and credit facilities, transport and communication systems, etc., are some important factors which have a great impact on the productivity of any industry.

1.4 Various Reasons for Low Industrial Productivity

Following reasons lead to low productivity:

- Ineffective use of resources instead of their optimum uses
- Persuance of non-productive/unnecessary activities
- Less productive labours
- Disputes among workers
- Poor information flow
- Excessive reworks resulting in the loss of productivity
- Wastage of materials
- Usage of old machines leading to frequent breakdowns and stoppage of production
- Extra long inventory

The researchers have come to the conclusion that ineffective use of available resources, non-productive or rewardless activities, and an inadequate and rather poor information flow are the three primary causes of low productivity in industries. The majority of researchers are of the opinion that dispute among workers can be one of the primary causes leading to low productivity.

1.5 Productivity Measures

In order to measure productivity of a nation or an industry, it is necessary to operationalise the same concept of productivity in all units or companies. The calculation of productivity of a nation or an industry is based on the time series of SNA, System of National Accounts. National accounting is a system based on the recommendations of the United Nations to measure total production and total income of a nation, and how they are utilised.

Productivity Measurement is one of the important functions of Industrial Engineering Departments in companies. In fact, it is the quantification of both, the output as well as input resources of a productive system. The intent is to come up with a quantified monitoring index. The goal of Productivity Measurement is to enhance productivity, which involves a combination of increased effectiveness and a better use of available resources. It facilitates planning and controlling productivity levels in the companies. The objectives of Productivity Measurement include better technology, efficiency, real cost-savings, benchmarking production processes and raising living standards.

Technology: A frequently stated objective of measuring Productivity Growth is to trace the technical change. Technology has been described as "the currently known ways of converting resources into outputs desired by the economy".

Efficiency: The quest for identifying changes in efficiency is conceptually different from identifying the technical changes. Full efficiency in the engineering sense means that a production process has achieved its maximum level of output physically achievable with current technology and a fixed amount of inputs.

Real cost-savings: Real cost-savings implies apragmatic way to describe the essence of measured productivity change. Although it is conceptually possible to isolate different types of efficiency and technical changes, and the economics of scale, it remains a difficult task to practice.

Benchmarking production processes: In the field of business economics, comparisons of productivity measures for specific production processes can help to identify inefficiencies.

Living standards: The measurement of productivity is a key element towards assessing the standard of living of people. A simple example is per capita income, probably being the most common measure of the living standards.

1.6 Models to Measure Productivity

Different models can be used to measure productivity. Some of them are as follows:

1. **Kendrick Creamer Model:**

   Kendrick and Creamer (1965) introduced productivity indice at the company level in their book, "Measuring Company Productivity". They proposed two types of indice.

   (*a*) **Total Productivity**: Total productivity index for the given period= (measured period output in base period price)/ (measured period input in base period price)

   Total factor productivity index = **Net output / Total factor input**

   Net output   – intermediate goods and services

   Total factor input – per man hour input and total capital

   (*b*) **Partial Productivity:** Partial productivity of labour, capital or material productivity index can be calculated as:

   Partial Productivity = (output in base period price) / (any one input in base period price)

2. **Craig – Harris Model:**

   Craig and Harris defined total productivity measure as:

   $$P_t = Qt / (L + C + R + Q)$$

Where;

P_t = Total productivity;

Q_t = Total output;

L = Labour input factor;

R = Raw material input factor

Q = Other miscellaneous goods and services input factor

The output is defined as the summation of all units produced times, their selling price, plus dividends from securities and interest from bonds and other such sources – all adjusted to base-period values.

3. American Productivity Centre Model:

American Production Centre has measured that expresses profitability as a product of productivity and price factor. It is done as follows:

Profitability = Sales/Cost

= (output quantity)*(Price)/ (input quantity) *(unit cost)

= (Productivity)* (Price Recovery Factor)

Where productivity = output quantity / input quantity

4. Sumanth's Total Productivity Model:

Total productivity (TPM) = Total Tangible Output / Total Tangible Input

Where:

Total Tangible Output = Value of finished units produced + value of partial units produced + dividends from securities + interest from bonds + other income

Total Tangible Input = Value of (human + material + capital + energy + other expenses) inputs used.

Sumanth provided a structure for finding productivity at product level and summing product-level productivities to total firm-level productivity. The model also has the structure for finding partial productivities at the product level and aggregating them to product-level productivities.

Summary

This chapter deals with productivity, its objectives and measurement models, etc. Productivity is simply the efficiency of a Production System. It is expressed as the ratio of outputs to inputs in production. Increasing national productivity can raise the living standards of people because more real income improves peoples' ability to purchase more goods and services, thereby improving the condition of housing, and education, and contributing more towards social, environmental and industrial growth programmes. Productivity is considered to be a key source of economic growth and competitiveness. There are various factors affecting productivity, viz., human, technical, managerial, natural, sociological, political and economic. The prime reasons for low industrial productivity are poor utilisation of resources, workers' disputes, excessive inventory, reworking and frequent mechanical breakdown, etc.

Productivity Measurement is the quantification of the output products as well as of the input resources of a Production System. The main objectives of Productivity Measurement include technology, efficiency, real cost-savings, benchmarking production processes and the living standards. Normally, Kendrick-Creamer, Craig Harris, American Productivity Centre and Sumanth Total Productivity Models, etc., are being used to measure industrial productivity. To conclude, in today's highly competitive global environment, Indian industries especially in the region of Uttrakhand, will have to select such advanced technology which may generate cost-efficient and good quality optimal outputs to ensure high industrial productivity.

❑❑❑

2

Introduction to Work-Study

2.1 Work-Study

Work-study may be defined as the study of a job for the purpose of finding the preferred method of doing it, and also determining the standard time taken to perform it by the preferred (or given) method. Work-study, therefore, consists of two areas of study: method study (motion study) and time study (work measurement).

2.1.1 Role of Work-Study In Improving Productivity

In order to understand the importance of Work-Study, we need to understand the importance of method study and that of time study as well .

Method study (also sometimes called Work-Method Design) is generally used to improve the method of doing work. It is also applicable to new jobs as well as the existing ones. When applied to existing jobs, method study aims to find better methods of doing those jobs that are economical, safe, require less human effort, and need shorter manufacturing time. The better method involves an optimum use of materials and right manpower so that work is performed in a well-organised manner, leading to increased resource utilisation, better quality output and lower costs.

It can, therefore, be stated that by method study we can have a systematic way of developing human resource effectiveness, providing high machine and equipment utilisation, and making economical use of materials in a safe manner.

Time-study, on the other hand, provides the standard time, that is the time required by a worker to complete a job by the standard method. Standard time for various jobs is necessary to estimate the following accurately:

- Manpower, machinery and equipment requirements
- Daily, weekly or monthly requirement of materials
- Production cost per unit as an input to improve making or buying decisions
- Labour cost
- Workers' efficiency and giving incentives to workers besides wage payments.

By the application of method study and time study in any organisation, we can thus achieve greater output at less cost, and of better quality; thereby achieving higher productivity.

2.1.2 Importance of Work-Study

Without measurements, there can be no management, and if the measurements are inaccurate there will be mismanagement. When it comes to measuring the standard times of various operations in the needle trade, work-study is a powerful tool. The prime value of work study lies in the fact that by carrying out its systematic procedures, a manager can achieve results as good or as better than the less systematic genius was able to do in the past.

Work-Study is a way to enhance the productivity of an industry by changing the method of work, by involving a very limited or no capital disbursement. Hence, it is done in a systematic way without leaving any aspect of production ignored. It is the most skilful way of investigating standards of performance as:

1. It ensures adequate usage and savings of various assets.

2. It is a means of enhancing the production efficiency (productivity) of the firm by eliminating the waste and unnecessary operations.

3. It is a technique to identify non-value adding operations by investigating all the factors affecting the job.

4. It is the only accurate and systematic procedure-oriented technique to establish time standards.

5. It is going to contribute to the profits as the savings will start happened immediately and continue throughout the life of the product.

6. It has a universal application. It is simply a tool which can be applied everywhere and used with success wherever work is having done or plant is having operated. It can be used not only in manufacturing units, but also in offices, stores, laboratories and service industries, such as wholesale and retail distribution, and on farms as well.

Work Study can contribute towards the improvement of working conditions at workplace by exposing hazardous operations and developing safer and better methods of performing operations.

2.2 Human Considerations In Work-Study

Success in application of any tools or techniques generally depends on the people who device them. Secondly, they are majorly influenced by the people on whom those techniques are applied. This holds water for the work study as well where the main focus is on the investigation of manual work, with an aim to economise human efforts. Investigation involves critical questions associated with the efficiency and effectiveness required for the work under consideration; this may cause frustration to the worker or the group of workers involved in the job. It is, therefore, necessary that right at the beginning the objective of conducting such a study be made clear to all concerned workers and supervisors by the management and the person doing work-study. In the first stage, the work-study person should intentionally record all relevant data concerning to the work. Successful implementation of any result mainly depends on the prevailing relationship among the workers and management. It is always easier to implement any change in an organisation where exists a mutual trust between the worker and management.

Therefore, human factors play a very significant role in the successful implementation of the objectives reached by work-study in a company. The management is responsible for spelling out the objectives and planning of the work. The supervisor translates these plans into day-to-day operations and continuously monitors them. The workers carry out the operations and the person responsible for work study carries out his task. Each of them has to contribute positively if the study has to succeed. The objective of this section is to look at the ways in which these different interacting groups can contribute.

2.3 Relationship of Work-Study Person with Management

As the management puts forward the objectives and plans of different activities, the supervisor's job is to translate these plans into daily operations, and oversee their progress by ensuring that the workers are giving their best to reach the desired performance level, and make relevant tools and find out available techniques. Therefore, the supervisor acts as the liaison officer between the management and worker. Now let us first discuss the roles of the management.

The role of management for successful application of work study can be summarised as follows:

1. The top management sometimes shows an indifferent attitude towards the idea put forward by work-study department; indeed, any effort to initiate work study is generally doomed. The management must clearly define the organisational goals and objectives. Otherwise, the workers and supervisors may set their own goals and objectives, which may be inconsistent with those of the management. Since work study intends to identify a different method for doing work, it ought to be defined with regards to the organisational goals.

2. The management strategy should be such that unproductive time gets minimised, which could be utilised as a part of the work study. The management should be ready to accept the suggestions and criticism as well.

3. The management should attempt to maintain a cordial relationship with the workers and provide a congenial work environment for them. This would help the management to build up a relationship of mutual trust which would prove for any project beneficial in the long run.

2.4 Relationship of Work Study Person and Supervisor

A supervisor or a foreman remains in direct contact with the workers. The success of anything introduced his shop floor largely depends upon the attitude of the foreman towards the newly introduced method. Moreover, it is essential to assess how much they are convinced about its benefits. The role of the supervisory staff in work study applications are summarised as follows:

1. In order to maintain perfect liasion between the management and workers the supervisor must clearly communicate the organisational objectives to the workers. At the same time, he should be able to portray a clear picture to the management about the practical problems on the shop floor, so as to enable the management to set realistic goals.

2. The work-study person who is much closer to the actual jobs at the shop floor than the management should be fully aware of the different aspects of the work and its limitations. This will help him identify potential areas of improvement, and he could be of great help to the management in selecting proper objectives for work study.

3. Since supervisor is responsible for executing plans, he has to be associated with the study right from the selection of the job, its analysis and implementation. This calls for an open mind. Status quo is usually a preferred choice of workers as well as the supervisor; yet the supervisor should be aware of this aspect of

human behavior that detests change. Hence, he should contribute to the study his observation by sharing his expertise on the work with the work-study man.

2.5 Relationship between Work-Study Man and Workers

A worker often fears that work study by increasing productive efficiency may lead either to retrenchment or relocation. To quell this fear the worker should also be given the same training that the regular work-study persons receive; thus he/she can take part with them in exploring potentials on the shop floor.

Worker plays a crucial role in the successful application of any study, since he/she is the actual person who would in reality perform the job on the shop floor. His/her ethics and attitude, his/her behavior as an individual as well as in group must be studied properly.

The workers should not ignore their job or waste time without any reason. As we have observed in the previous chapter, due to the negligence of the workers the time spent on the shop floor may be unproductive at times. Hence, they must realise the fact that this resulting lower productivity will affect their carriers in the long run.

The workers should take interest and initiative in their work, and in their work related functions. Some time, it is possible to select the jobs by observing the interest of the worker. Individual, formal or informal groups or sometimes unions can be used as a platform to initiate the study.

2.6 The Work-Study Person

While the management, supervisors and workers are insiders, so far as the information is concerned, the work-study person is an outsider. His/her job thus becomes the most complicated in terms of coordinating with all three groups, and in trying to bring about changes in the system. The following are some of the hints that should be understood by him/her to act rationally:

1. Enhancing productivity should be dealt in an impartial way without much emphasis being placed on the productivity of

labour. In most of the enterprises, in developing countries and in industrialised countries, there is a big scope to increase the productivity by applying the suggestions of work-study person to improve plant utilisation and operations to make more effective use of space, and to secure greater economy of materials before question of increasing the productivity of the labour force arises. The importance of studying the productivity of all the resources of the enterprise, and not confining the application of work-study to the productivity of labour alone, cannot be over emphasised. It is very natural that workers would dislike the efforts being made to improve their efficiency while they can see glaring inefficiency on the part of the management.

2. The person must be fully willing to give his/her best for the purpose of study. He/She should be very honest, frank and transparent; he/she should not attempt to hide the changes envisaged. He/She should frankly answer the questions and provide information obtained from his/her studies. Work-study, honestly applied, has nothing to hide.

3. Seniors must be given comprehensive details regarding the study with suitable reasons in order to adopt the improved and more suitable methods and techniques.

4. Interacting ideas and views with workers makes them feel an integeral part of the organisation.

5. It is important that person doing work-study must keep in his/her mind that a mere enhancement in productivity is not enough. The human factor can not be ignored; job satisfaction is paramount. The work-study person should deal with the issue of productivity by considering the ways to make jobs more satisfying, involving less fatigue.

2.7 Attributes of a Work-Study Person

The need to translate concepts and ideas into reality calls for certain qualifications. These can be summarised as follows:

1. **Education:** The person taking charge of the work-study must at least have good secondary education as an equivalent school-leaving certificate. It is unlikely that anyone with much education will be unable to grasp things during the work-study course, though there may be with a few exceptions. However, if the work-study person is also involved in observing other production management problems, a university degree in engineering or management or an equivalent certificate becomes an important requirement.

2. **Experience:** It is mandatory for applying candidates to have practical experience in their respective industries. This will facilitate them to face various challenges while dealing with the labour force.

3. **Personal qualities:** He must be able to come up with fundamental systems and tools which may save time and effort, enhance cooperation among engineers and technicians in advancing these techniques. Nevertheless, one who is good at these things may not always be clueful in human relations. In some large companies the methods department is separate from the work measurements department, though they may he under the same head. The following are the essential attributes required:

   (*a*) **Sincerity and Integrity:** The work-study person should be sincere and honest in order to win the confidence and respect of people he/she is working with.

   (*b*) **Enthusiasm and motivation:** He/She must be fully interested in the job, giving importance to what is having done, making others feel his/her positivity and dedication.

   (*c*) **Diplomacy and Negotiation:** Efficient skills of dealing with people come by understanding people, respecting their feelings with generous and thoughtful words.

   (*d*) **Self confidence:** This can only be achieved with proper training and incorporating work-study efficiently and successfully in order to win respect from his seniors as well as the top management.

Summary

This chapter deals with the introduction to work study. It mainly targets the human considerations in the work-study. It also explains the role of management, work-study person, supervisors and workers in the execution of the work-study programmes initiated by the organisation. Besides, it also highlights the qualities of a work-study person.

3

Method Study

Introduction

Method study, aims to achieve better ways of completing a work; and for this reason method study is sometimes called Work-Method Design. As a productivity improvement step, Method Study helps to produce the same output using fewer resources, or to produce more with proportionately less inputs. It reduces, if not eliminates the waste completely. Method Study ensures creativity, innovativeness, optimal decision-making power, good organisational practices and better communication. Here are some symptoms whose presence can warrant the need of Method Study:

Given below are some shortcomings in a workplace that warrant the need of Methods Study.

1. Dissatisfaction among the clients/beneficiaries
2. Increased operating costs
3. Low morale of the work force
4. Higher wastage of materials, machinery, labour, space and services
5. Excessive movement and backtracking,handling of materials, men and motion
6. Presence of bottlenecks in production
7. More overtime done by workers
8. More rejections and reworks

9. Low quality issues

10. Poor working conditions, etc.

3.1 Definition of Method Study

Method study can be defined as the procedure of recording, analysing and examining of existing or proposed method of doing work systematically for the purpose of improvement and development, and application of more effective method.

Objectives: The important objectives of work study are as follows:

1. Improvement in processes and procedures employed

2. Improvement in factory and workplace layout

3. Improvement in the design of the plant and equipments

4. Reduction of unnecessary fatigue caused to workers

5. Use of improved materials, machines and manpower

6. Better working conditions

3.2 Procedure of Method Study

The following general steps describe the procedure for making a Method Study.

1. Select the job

2. Record the information

3. Examine the information critically

4. Develop the most practical, economical and effective method

5. Install the new method as a standard practice

6. Maintain the standard practice by regular follow up

3.2.1 Selection of Job for Method Study

An engineer should always attempt to select those jobs for improvement which are unpopular among the employees, or are considered "unworthy". By improving the method of such work, he/she would earn goodwill of the employees as well as of the management, and can expect their cooperation in the future. While selecting job considerations should be given to the following factors:

- Economic,
- Technical, and
- Human Factors.

3.2.1.1 Economic Factors

If the economic importance of a work is found to be insignificant,it is adviceable to start or continue a Method Study on it. Priorities should be given to those jobs which offer greater possibilities for cost reduction. Such jobs can be easily identified, as they have:

- Have high labour cost
- Consume more time
- Are performed mechanically where less manpower is required
- Have higher frequency of occurrence, *i.e.,* they have larger demands
- Have bottlenecks in production line
- Prone to higher proportion of accidents
- Require more material or men-movement over long distances
- Incur high scraps and reworks costs
- Require higher payment to workers because of overtime

3.2.1.2 Technical Factors

The method-study engineer must have the necessary technical knowledge about the job to be studied. To illustrate, consider that a particular machine tool is proving to be a bottleneck. The output from this machine is not reaching the assembly line in the required quantity and quality. Through a preliminary study, it is found that it is running at a lower speed than that recommended for the work, and tool materials used. More increase in speed may not be the solution to this problem. It may be possible that the machine itself is very rigid to operate at a higher speed or take a deeper cut. Mere increase in speed may increase the output but its quality may be seriously affected.

3.2.1.3 Human Factors

Emotional attitude of the workers to the method study that brings about changes in their working are also an important consideration. If the study of a particular job is suspected to cause unrest or ill feeling, it should not be undertaken, however useful it may be from the economic point of view. It is always better to take up first those jobs which are considered unclean, unsafe, unpleasant, tedious, or highly timing, and improvements should be brought about in them as a result of method study.

3.2.2 Information Collection and Recording

3.2.2.1 Information Collection Techniques

The accuracy of data about the problem to be studied is important for the development of new or improved method. The following techniques are used for the collection of information / data about the task under consideration. They are not exclusive for any particular method study problem, infact some or all the techniques may be employed at most of the places .

Observation: It is a common technique used for collecting information about the present method or the existing problem. The method-study person visits the site where the work is currently being carried and observes various steps in the method of work being followed.

Discussion: Discussion with the task force, or with those who supervise the work can frequently provide information which is not obtainable by observation. The discussion technique is commonly used where irregular work is involved or where one is trying to analyse past work in order to improve efficiency in future.

Records: Valuable information can be obtained from past records concerning production, cost, time, inventory and sub-contracts. For certain type of information concerning the past practices, sometimes this is the only way to get authentic data.

Motion Pictures or Video Films: Accurate and most detailed information can be obtained by taking video films. Information obtained by this procedure can easily be forwarded to management, if needed. The film can be used to focus attention at particular point or motion in an operation.

3.2.2.2 Information Recording Techniques

There are three main types of information recording techniques. These are as under:

(*a*) Process Charts

(*b*) Diagrams

(*c*) Templates

(*a*) **Process Chart:** A **Process Chart** is a graphical representation of the activities that occur during manufacturing or servicing job. There are several types of process charts. They can be divided into two groups.

  (*i*) Those which are used to record a process sequence (*i.e.,* series of events in the order in which they occur) but do not depict the events according to the time scale. Charts under in this group are:

  (*a*) Operation process chart

  (*b*) Flow process chart – (man / material / equipment type)

  (*c*) Operator chart (also called Two Handed Process chart)

  (*ii*) Those which record events in the sequence in which they occur on a time scale so that the interaction of related events can be more easily studied. Charts that fall in this group are:

  (*a*) Multiple Activity chart

  (*b*) SIMO chart

(*b*) **Diagrams:** A diagram gives a pictorial view of the layout of the workplace or shopfloor on which locations of different machines, equipments, etc., are shown. The movement of man or material is then indicated on the diagram by a line or a string.

The diagrams highlights the movement so that analyst can take steps to simplify or reduce it and thus affect saving in time or reduction in collisions / accidents/ time.

(*i*) Flow Diagram

(*ii*) String Diagram

Two types of diagrams are common: Flow diagram and string diagram.

(*i*) **Flow Diagram:** The term flow diagram is used in theory and practice with different meanings. Most commonly the flow charts and flow diagrams are used in an interchangeable way to represent a process. For example, the two separate definitions are as follows:

Flow diagram visually displays interrelated information, such as events, steps in a process, and functions in an organised fashion, such as sequentially or chronologically.

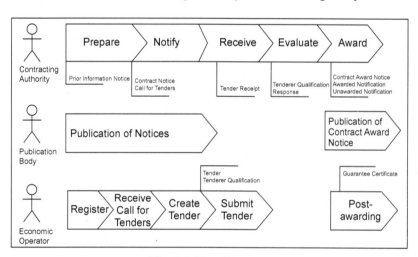

Fig. 3.1 Flow Diagram

Flow diagram is a graphic representation of the physical route or flow of people, materials, paperwork, vehicles, or communication associated with a process, procedure plan, or investigation. Figure 3.1 shows an example of flow diagram of tending process.

Though, flow process chart indicates the sequence of events, it does not show the movements of men, machine, materials, etc., while the work is being accomplished. Given below are steps to construct flow diagrams:

(*a*) Draw a scale plan of work area.

(*b*) Mark the relative position of machines tools, work bench, store, racks, etc.

(*c*) From various observations, draw the actual movements of the materials or workers on the diagram and indicate the direction of movement by arrow heads.

(*ii*) **String Diagram:** When the paths are many and repetitive, a flow diagram becomes congested and not understandable. Under such a situation a string diagram becomes a preferred option. String diagram is one of the most useful and simplest technique of method study. It can be defined as a scale model on which a thread is used to trace the path or movements of men and materials during a specified sequence of events. It can also be stated that string diagram is a special form of flow diagram. As a thread is used to measure distance, it is necessary that the string diagram should be drawn up to a scale; though the same is not necessary in the case of flow diagram.

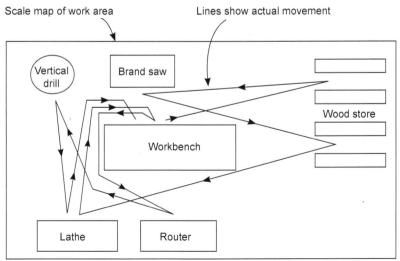

Fig. 3.2 String Diagram

Here are the steps to help one construct a string diagram.

(*a*) Draw a scaled layout of the working area.

(*b*) Mark the various features, like machines, work benches, store, etc.

(*c*) Place scaled diagram on a soft board and strike pins at all the places which form the path of worker and materials.

(*d*) A continuous coloured unstretchable string, taken from the start to last pin, is wound to mark the path followed by men or material.

A thread should be rounded by each pin to measure the movements. The thread when measured gives approximately the total distance travelled by a worker or material. A typical string diagram is shown in Figure 3.2.

(*c*) **Templates and 3-D models:** Instead the scale plan of shop/ section their templates and three dimensional model give more realistic view of a workplace. The bottlenecks, congestion and back tracking can be better visualised

Templates: A two-dimensional scaled cut-outs made from thin card sheet representing machinery, equipments, workbenches, fittings, furniture, etc., can be used for developing new layouts and methods. The templates sometimes may have pieces of permanent magnet attached to them, so that when used on iron boards they would remain glued is the place. They show the actual floor space utilisation. Templates saves lot of time and human effort which otherwise would be spent in making drawing for each alternative layout. They present various visual characteristics, advantages and limitations of a plant/shop layout.

Advantages of two Dimensional Templates are as follows:

1. They are less expensive.

2. They can be easily interpreted.

3. There number of copies can be made.

4. Time taken to make a template is less as compared to the time taken to prepare a 3D model.

Limitations:

1. Only technical persons can interpret it correctly.

2. Overhead facilities/ equipments cannot be visualised.

Three Dimensional Model: They are scaled 3-D model of facilities and are more near to the real picture. They help in easy understanding of lighting, ventilation, maintenance and safety aspects that may be important in a method along with the position of machines ,equipments and movements. Such models are often of great value in demonstrating the advantages of the proposed changes to all concerned. However, their use is limited because of higher cost involved. Some computer softwares are available nowadays which help in constructing the layouts and fecilitate the possibility of visualising the working of process in a systematic way.

Advantages of 3D Models:

1. Layout is very easy to understand.

2. Models can be shifted easily and quickly to study the operations.

3. Overhead structure and equipment can be visualised.

Limitations:

1. They require more space as compared to templates.

2. They are more expensive.

Before taking up descriptions of these charts or diagrams, it is necessary to know the various elements of work.

3.2.2.3 Elements of Work

There are five basic elements of work: Operation, Inspection, Transportation, Delay, and Storage. Table 3.1 gives the description and symbols by which all work elements are represented.

Table 3.1 Basic Work Elements

Name	Symbol	Description
Process	▭	Process action step
Flow line	⟶	Direction of process flow
Start/terminator	⬭	Start or end point of process flow
Decision	◇	Represents a decision making point
Connector	·◯	Inspector point
Inventory	△	Raw material storage
Inventory	▽	Finished goods storage
Preparation	⬡	Initial setup and other preparation steps before start of process flow
Alternate process	▢	Shows a flow which is an alternative to normal flow
Flow line (dashed)	– – – ⟶	Alternative flow direction of information flow

Sometimes, more than one element occur simultaneously. It is shown as combined element with combined symbol. Examples are, "Operation in combination with inspection", and "Inspection in combination with Transportation".

Operation Process Chart: An operation process chart gives the chronological sequence of all work elements that occur in a manufacturing or service process. It also shows materials used and the time taken by operator for completing different elements of work. Generally a process chart is made for full assembly, that is, it shows all the operations and inspections that occur from the arrival of raw material to the packaging of the finished product. Figure 3.3 shows the operation Process Chart.

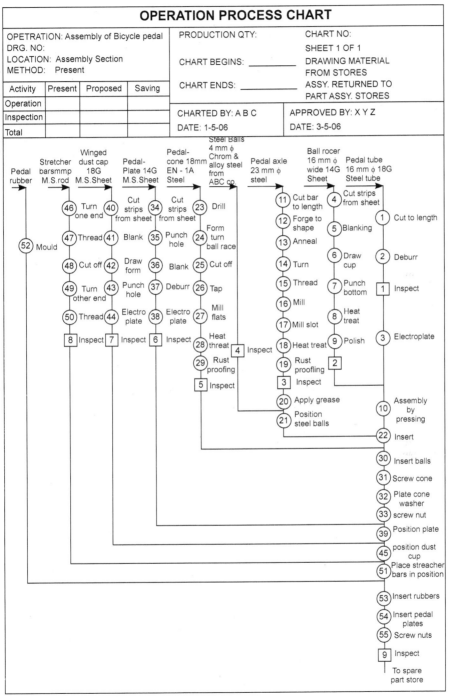

Operation process chart for assembly of bicycle pedal

Fig. 3.3 Operation Process Chart

Operation process chart the following advantages.

(*a*) It helps in visualising the full process so that necessary improvements may be made, if required.

(*b*) It highlights the relationship between different activities.

(*c*) In this chart mainly operations and inspections are considered, i.e., it makes use of two symbols only.

(*d*) Descriptions of operations and inspections are written on the right-hand side of the symbols in the chart.

Flow Process Chart: A flow process chart is used for recording more details than an operation Process Chart. It is made for each component of an assembly individually rather than for the whole assembly. A flow process chart shows the complete process and activities in terms of all elements of work. There are three main types of flow charts: material, operator and machine flow process chart. The product flow chart records the details of the events/activities that occur to a product, while the operator flow chart records the details how a person performs an operational sequence, i.e., what an operator does. The Machine Process Chart records what happens to the material, i.e., the changes the material undergoes in a particular condition or location. The figure 3.4 shows a flow process chart of a nuclear power productive hidden costs, such as delays, temporary storages, unnecessary plant. An important and valuable feature of this chart is its recording of

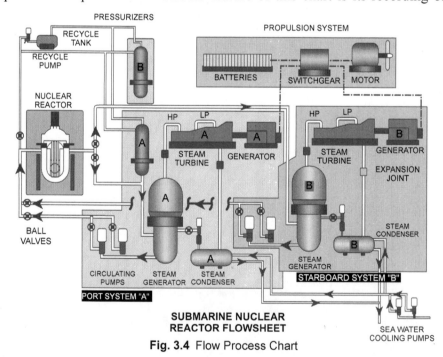

**SUBMARINE NUCLEAR
REACTOR FLOWSHEET**

Fig. 3.4 Flow Process Chart

non-inspections and unwarrented long-distances travels. When the time spent on these non-productive activities is highlighted, analyst can take steps to minimise it and thus reduce costs.

Operator Process Chart: It is also known as Left Hand – Right Hand chart and shows the activities of hands of the operator while he/she is performing a task. It mainly uses four elements of work: Operation, Delay, Move and Hold. Its main advantage is that it highlights un-productive elements, such as unnecessary delay, and holds it so that the analyst can take corrective action to eliminate or reduce them. Figure 3.5 shows a Operator Process chart.

Operator Process Chart

Chart No	Sht No:		Summary						
Drg. No:	Part No:			Present		Proposed		Savings	
Operation: Assembly of Electrical Tester				L.H	R.H	L.H	R.H	L.H	R.H
Location: Assembly Section		Operation		2	9				
Date: 10.02.05	Method: Present	Transport		3	8				
Operator: OPSharma	Analyst: KDSingh	Hold		1	0				
Workplace Layout:		Delay		0	2				
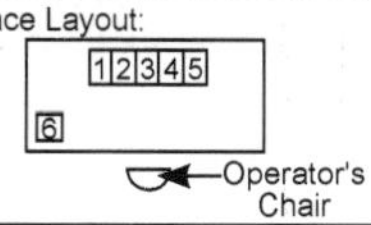 1 2 3 4 5		Time (min)		.50					
6		Part Sketch of Jig etc							
◁ Operator's Chair									

Left Hand Description	O ⇨ Q D	O ⇨ Q D	Right Hand Description
To barrel in bin 1			To filament in bin 5
Pick up barrel			Pick up
To work position			To barrel
Hold			Assemble in barrel
			To resistor in bin 4
			Pick up
			To barrel
			Assemble in barrel
			To clip in bin 3
			Pick up
			To barrel
			Assemble in barrel
			To cap in bin 2
			Pick up
			To barrel
			Position on clip
Hold			Screw in the barret
To bin 6			Rest on table
Release			Rest on table

Operator process chart for assembly of an electric tester

Fig. 3.5 Operator process chart

Multiple Activity Chart: Where a number of workers work in groups, or an individual worker handles two or more number of machines, their activities have to be coordinated for achieving better results. A multiple activity chart records the activities of all worker and machines on a common time scale simultaneously.

Worker-Machine process chartand Gang process chart fall in the category of multiple activity charts. A Worker-Machine chart is used for recording and analysing the working relationship between the operator and machine on which he works. It is drawn on time scale. Analysis of the chart can help in better utilisation of time by both, the worker as well as machine. The possibility of one worker attending more than one machine is also sought from the use of this chart. Figure 3.6 (a) and (b) show a multiple activity chart.

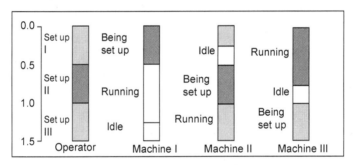

Fig. 3.6(a) Man-Machine Chart

Machine		Process		Process	
Operator A	Load		Un-load	Load		Un-load
Operator B	Pack			Pack		
Operator C		Store			Store	

Time (minutes) 0 2 4 6 8 10 12 14 16 18

Fig. 3.6 (b) Man –Machine Type Multiple Activity Chart

The purpose of multiple activity charts are:

(*a*) To findout the idle time being enforced on worker and machine.

(*b*) To optimise the work distribution between worker and machines.

(*c*) To balance the work team.

(*d*) To find out optimum number of workers in a group.

(*e*) Finally to develop an improved method of performing the task effectively.

Gang process chart: It is another type of multiple activity chart. A gang process chart is similar to worker-machine chart, and is used when several workers operate one machine. The chart helps in exploring the possibility of reducing both the operation time and idle machine time and to improve the efficiency of the gang operations.

Simultaneous-Motion (SIMO) Chart: SIMO chart is commonly used for micro motion-analysis of a small cycle repetitive jobs, high skills jobs and finds application in job components, such as assembly line, packaging and inspection. ASIMO Chartis another Left-Hand Right-Hand chart with the difference that it is drawn to time scale and in terms of basic motions called Therbligs. It is used when the work-cycle is highly repetitive and lasts for a very short duration. Figure 3.7 shows a template of SIMO chart.

SIMO Chart

Operation :							
Part drawing No. :				Chart No. :			
Method : .... Present/Proposed			Date :			
Operation No. :			Charted by :			

Work counter Reading	Left Hand Description	Therbligs	Time	Time in 200/m	Time	Therbligs	Left Hand Description

Fig. 3.7 Template of SIMO Chart

Constructing SIMO Chart: A video film or a motion picture film is shot of the operation as it is carried out by the operator. The film is analysed frame by frame. For the left hand, the sequence of Therbligs (or group of Therbligs) with their time values are recorded on the column corresponding to the left hand. The symbols are added against the length of column representing the duration of the group of Therbligs. The procedure is repeated for the right hand and other body members (if any) involved in carrying out the operation. Figure 3.8 shows an example of SIMO Chart

SIMO CHART

Operation: Assemble Clamps and Bolt			Part: CC-5		Summary	Left Hand	Right Hand
Operator Name and No.: P.B.#120					Effective Time:	3.3	11.4
Analyst: P.Kumar			Date:08/08/05		Ineffective Time:	10.2	2.1
Method: Present Proposed					Cycle Time: = 13.50 sec.		

Sketch of workplace: U Bolt, Nuts, Clamps, Assembled Units, Main, 35 cm

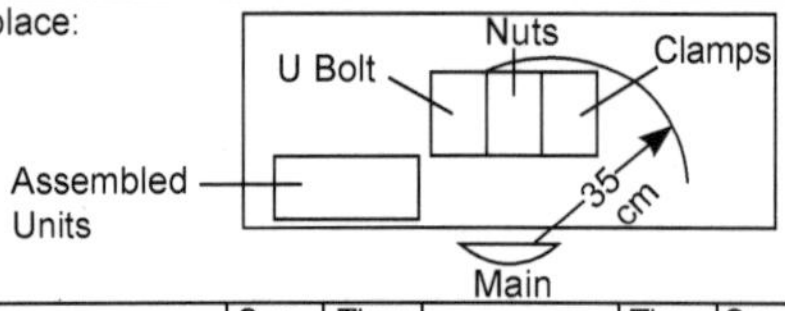

Left hand Description	Sym-bol	Time Sec		Time Sec	Sym-bol	Right hand Description
Get U Bolt	RE G	1.00		1.00	RE G	Get Clamp
Place U Bolt	M P RL	1.20		1.20	M P RL	Place Clamp
				1.00	RE G	Get Nut #1
				1.20	M P	Place Nut #1
				3.00	U / RL	Run Down Nut #1
Hold U Bolt	H	10.20		1.00	RE G	Gut Nut #2
				1.20	M P	Place Nut #2
				3.00	U / RL	Run Down Nut #2
Dispose off Assembly	M RL	1.10		0.90	UD	Wait

Fig. 3.8. SIMO Chart

Applications of SIMO Chart: From the analysis shown about the motions of the two hands (or other body members) involved in doing an operation, inefficient motion pattern can be identified and any violation of the principle of motion-economy can be easily noticed. The chart, therefore, helps in improving the method of doing an operation so that balanced two-handed actions with coordinated foot and eye motions can be achieved, and ineffective motions can either be reduced or eliminated. The result is a smoother, more rhythmic work-cycle that keeps both delays and operator fatigue to the minimum extent.

3.2.3 Critical Examination

Critical examination of the information recorded about the process in the form of charts / diagrams is a very important component of the method study. In this, each element of the work, as presently being done and recorded on the chart is subjected to a systematic and progressive series of questions with the purpose of determining reasons for which it is done. This examination thus requires exhaustive teamwork where everyone's contribution can prove useful, besides optimum use of all available sources related to technical information.

For most common procedure of carrying out critical examination we use two sets of questions: Primary questions, and Secondary questions. Selection of the best way of doing each activity is later determined to develop new method which is introduced as a standard practice method. A general-purpose set of primary and secondary questions is given below:

3.2.3.1 Primary Questions

These are used to indicate the facts and reasons underlying them.

Purpose. The need of carrying out the activity is challenged by the questions:

What is achieved?

Is it necessary?

Why?

Means. The means of carrying out the activity are challenged by the questions:

How is it done?

Why that way?

Place. The location of carrying out the activity is challenged by the questions:

Where is it done?

Why there?

Sequence. The time of carrying out the activity is challenged by the questions:

When is it done?

Why then?

Person. The level of skill and experience of the person performing the activity is challenged by the questions:

Who does it?

Why that person only?

The main purpose of the primary questions is to make sure that the reasons for every aspect of the presently used method are clearly understood. The answers to these questions should clearly bring out any part of the work which is unnecessary or inefficient in respect of means, sequence, person or place.

3.2.3.2 Secondary Questions

The aim of secondary questions is to arrive at appropriate alternatives and consequently at the mean of improvement to the presently used method:

Purpose. If the answer to the primary question 'Is the activity necessary"?is convincingly 'Yes', alternatives to achieve the object of carrying out the activity are considered by the question:

'What else could be done'?

Means. All the alternative means to achieve the object are considered by the question:

'How else could it be done'?

Place. Other places for carrying out the activity are considered by the question:

'Where else could it be done'?

Sequence. The secondary question asked under this heading is:

'When else could it be clone'?

Person. The possibilities for carrying out the activity by some other persons are considered by asking the question:

'Who else should do it'?

This examination involves the search of alternative possible method within the imposed restrictions of cost, volume of production, and similar other variables. The answers to the following questions are then sought through evaluation of the alternatives methods:

What should be done?

How should it be done?

Where should it be done?

When should it be done?

Who should do it?

These answers are the basis of the proposals for the improved method. The evaluation phase requires the work-study engineer to consider all the possibilities with respect to the four factors—economic, safety, work quality and human—the economic factors being the most important in most of the situations.

Economic considerations to any alternative means to determine 'How much will it cost'? and 'How much will it save'? Similarly, the purpose of evaluating safety factor is to ensure that the alternative selected method shall not make the work unsafe. The evaluation of quality factor determines whether the selected alternative method shall make for better quality of the product or better quality control on the process. And finally the human factor considerations shall ensure that the new selected method is more interesting, easy to learn, safe, less monotonous and less fatiguing to the operator/worker.

Table 3.2 shows a sample sheet used for critical examination the use of which can be quite helpful in method study.

Table 3.2 A sample sheet for critical examination

	Primary Question		Secondary Questions	
Sr. No.	Facts	Challenge/Why?	Alternatives	Proposed Improvement

3.2.4 Developing the New/Better Method

Steps involved in developing the better method than the existing ones are:

(*i*) Record the existing method through a critical examination.

(*ii*) Consult all level of management and workers for their suggestions and objections.

(*iii*) Test the modified proposed method.

(*iv*) Prepare and put up the report on proposed method before the management for its final approval and implementation.

While developing the new/better alternative method for doing a task the following points may be considered:

(*i*) Where and how to use 'man' in the process?

(*ii*) What better work procedure can be adopted?

(*iii*) What better machines/equipment can be used?

(*iv*) What better layout of work station, shop or factory can be used?

In deciding whether a particular element of work (operation, inspection, or transportation) should be carried out manually or with the help of a device, method-study engineer must be well aware of workers things which cannot do, or do in an inferior than machine requires.

For instance:

(*i*) Exert amount of force, as required in metal-cutting.

(*ii*) Do high speed computations of complex nature problems.

(*iii*) Perform repetitive tasks without suffering from side effects like boredom and fatigue.

(*iv*) Move at high speeds for hours continuously.

(*v*) Perform several tasks simultaneously.

(*vi*) Perform satisfactorily in an environment where conditions relating to cold, heat, noise, dampness, etc., are extreme.

In contrast, machines prove inferior generally when for carrying out a task it is necessary to

(*a*) Think creatively or inductively

(*b*) Learn

(*c*) Decision

(*d*) Generalise

In most cases, the relative roles of man and machine vary from one extreme end in which entire process is manual to the other extreme in which the process is completely mechanised with the presence of man only for monitoring and taking care of trouble-shooting, maintenance, etc.

The various alternative methods of carrying out essential elements of work, method study engineer has now to choose the best alternative method. He decides upon the criteria, which may fix costs involved, running cost, production rate, operator's fatigue, operator's learning time, and similar other things. The weight to each criterion is fixed and performance of each alternative is predicted with respect to each criterion. The one which gets the maximum points is selected for adoption as a new and better method. Detailed specifications of this method are prepared with the description of procedure, workplace-layout and materials/ equipments to be used. It is important to communicate the proposed work method to those responsible for its approval, similarly it is equally important to communicate to those concerned with its installation. For example, instructors and supervisors who are actually responsible to instruct operators and set up the machinery and workplace layouts.

3.2.5 Installation of Improved Method

When the proposals of the improved/better method for a job are approved by the management of the company, the next step is to put this method into practice, i.e. its installation. Installation of method requires necessary prior preparation for which the active support of everyone concerned is essential.

The step taken to install the new method include the following points:

1. The first step is to gain acceptance from the workers and their representatives involved. The method change may affect the routine of workers, may require more paper work for changed wages, costs and planning. It would mean extra work for the purchase department. It may require relocation of staff from one section/shop to another.

2. Retraining the workers. The extent to which workers need retraining will depend on the nature of job and changes involved in the new alternative method. It is much more for those jobs which have a high degree of manual dexterity and where the workers have been doing the work by conventional methods. The use of films demonstrating the advantages of new method as compared to conventional one are often very useful in retraining the workers.

3. In order to arrange the requirements of the new method the following steps need to be taken.

    (*i*) Arranging the necessary tools, jigs, fixture and equipments at all workplaces,

    (*ii*) Arranging building-up of necessary stocks of new raw materials, and running down of old stocks,

    (*iii*) Checking up the availability and continuity of all supplies and services, and

    (*iv*) Arranging any office records which may be required for the purpose of control and comparison of the method.

4. Taking other necessary actions. For example, if changes in working hours are required necessary instructions may be passed on to auxiliary services, such as electricity supply, transport, canteen, water supply. If change in wages is involved, information

about the date of installation must reach the accounts department well in time. Necessary instructions may be issued to every one concerned about the time table for the installation of the alternative method.

5. Perform a trial run to the new method. It is important from the view point of rehearsal of all the departments that are affected by the change of method. It is often advantageous to conduct the rehearsal while the old method is still operating, employees that both the methods would work simultaneously. The suggestions for minor variations in the proposed method, if they are worthwhile and cost effective, should be accepted and incorporated.

It is obvious that the method analyst has to be extra tactful and keep control throughout the period of installation. The installation is complete when the new method starts running smoothly.

3.2.6 Maintain the Standard Practice by Regular Follow Up

The work of the person conducting method study is not complete with the installation of the improved method; the maintenance of the new improved method in its specified form is also a part of his activities. The main aim of maintenance of the new improved method is to ensure that the workers do not follow the old method.

In order to maintain the new method effectively it is important to define and specify it clearly. An operator chart giving sufficient details of the tools, equipments, and workplace-layout, and operator-motion pattern is often very helpful.

The workers have a tendency to drift away from any new method implemented. The purpose of the method-maintenance is to check this tendency. But if it is found that the change from the method specified is in fact an improvement which can be made, it should be officially incorporated.

3.3 Therbligs

On analysing the result of several motion studies conducted, Gilbreths concluded that any work can be done by using a combination of some or all of 18 basic motions called Therbligs (Gilbreth spelled backward).

These can be classified as effective Therbligs and ineffective Therbligs. Effective Therbligs take the work progress towards completion. Attempts can be made to shorten them but they cannot be eliminated. Ineffective Therbligs do not advance the progress of work, and therefore attempts should be made to eliminate them by applying the Principles of Motion Economy. The following table lists the Therbligs, along with their mnemonic symbols and standard colours for charting.

Table 3.3 Therbligs Chart

Therblig	Color	Symbol/Icon	Therblig	Color	Symbol/Icon
Search	Black		Use	Purple	
Find	Gray		Disassemble	Violet, Light	
Select	Light Gray		Inspect	Burnt Orange	
Grasp	Lake Red		Pre-Position	Sky Blue	
*Hold	Gold Ochre		Release Load	Carmine Red	
Transport Loaded	Green		Unavoidable Delay	Yellow Ochre	
Transport Empty	Olive Green		Avoidable Delay	Lemon Yellow	
Position	Blue		Plan	Brown	
Assemble	Violet, Heavy		Rest for overcoming fatigue	Orange	

Search: (an eye turned)

The Search motion starts when the eyes and/or hands start to seek the object needed, and ends just as the object is located. The Gilbreths stated that in a search, "....the time and attention required...varies with the number of dimensions in which the search is performed." A single dimensional search might be locating a piece of paper on a desktop. A two dimensional search might be finding a light switch on a wall, and a three dimensional search would be locating a hanging pull-chain for a light or fan.

Find: (an eye looking straight)

If there is an puzzle in the Therblig system, find it. Dr. Barnes eliminated this Therblig, explaining that it was a mental reaction, at the end of the Search cycle. While other mental processes are included as Therbligs, this one is so momentary that the time taken for the find-function would be hardly worth-measuring. So, it should be kept available, since it may become important in a future application of the system.

Select: (an arrow aimed at an object)

This Therblig may be considered a part of Search. However, through usage by the Gilbreths, it was found to indicate locating an object from a group of similar objects. For example, an artist may search for a box of coloured pencils and then select the proper colour. The important thing to remember is that the search, find and select Therbligs may or may not be separate elements, depending entirely on the type of work being analysed.

Grasp: (a hand poised over an object)

In simplest terms, grasp is when the worker's hand grabs the object. The Therblig ends when the next Therblig, of use or transport-loaded, begins. There are actually many aspects to Grasp, which Gilbreths recognised and which continue to develop today. In this Therblig, the time taken is directly proportional to the ease of the grasp. For example, the more dimensions the object has, the quicker it can be effectively grasped. Frank Gilbreth observed that a sales clerk would put a slight crease in a cash register receipt so it rises above the counter surface, making it easier to pick up. An important element in saving time was whether the initial grasping of an object would be the proper grasp for the use or assemble function. In this respect, grasp has a close relationship with the position and pre-position Therbligs.

Hold: (a horseshoe magnet holding a bar)

Dr. Barnes said this Therblig was "....the retention of an object after it has been grasped, (with) no movement of the object taking place." To clarify, we can call Hold a grasp, of an object, occurring in one hand, while the other hand performs a Use or assemble function. While the Gilbreths considered this part of grasp, Mogensen and Barnes were correct in making it a separate Therblig, so as to alert the user to a negative Therblig, which should be eliminated. This is particularly true in using Therbligs in ergonomics, where static holding is an undesirable posture. By eliminating static Holding, you not only free up a hand for other uses, but also reduce overall fatigue.

Transport Loaded: (a hand-cupped, holding an object)

This Therblig begins after Grasp where the hand is doing "work" by moving the weight of an object, and ends just before the release load, use or assemble Therbligs. The main objective of this Therblig is to reduce the distance and subsequent time involved in transportation. However, an obscure note in the Gilbreth papers has even more important ramifications in applying Therbligs to Ergonomics. Distances and effort can be reduced by the old Gilbreth maxim of making gravity work for you, by having sloped bins. This type of storage bin would also improve the search function, since it would be easier to see the olyids.

Transport Empty: (an empty hand)

This is the motion of moving the unloaded hand from the point of Release Load to the next function within the sequence. It can also be considered as the hand motions involved between Select and Grasp, where the eye identifies the object and the hand moves towards it to grasp. This Therblig is a non-productive one, and as such, should be kept to a minimum.

Position: (an object, such as a pen, being placed in the hand)

This motion is the act of placing the object in the proper orientation for use. For example, a screw lies on the workbench in a horizontal orientation, but is to be used in a vertical position. Positioning would occur when the screw is picked up and rotated into the vertical position for inserting it into an object. This function may be completed during Transport Loaded or be a totally separate Therblig.

Assemble: (several items (lines) placed together)

This Therblig starts when two or more parts are placed together (a peg into a hole) and ends when either the assembled object is Transport Loaded or when the hand reaches for another part (transport empty). Long lengths of time for this Therblig open numerous possibilities for improvement. For example, in the case of placing a peg in a hole (each of the same diameter), both Gilbreth and Barnes found that you can speed assembly by increasing the size of the target. In the case of the peg, assembly time will be significantly shorter if the holes are countersunk, which aids in guiding the peg into the hole.

Use: (simple the letter U---for Use)

This Therblig should not be confused with assemble. Use it when an object is being operated as it was intended, and typically denotes a tool. For example, we would assemble a drill by placing it a bit in the chuck and tightening it, but we Use the drill to bore holes. Operation of controls on a machine would also be considered Use.

Disassemble: (Assemble symbol with one part removed)

This motion is essentially the opposite of assemble, depending on the circumstances. While it could be used where a mistake was made in Assemble, it could also be the act of removing a part from a jig or clamp, which held the part during the Use or Assemble motion.

Inspect: (a magnifying glass)

This Therblig involves the act of comparing the object with a predetermined standard. This act can employ one or all human senses, depending on the object and the desirable attributes being checked. The inspection can be for quantity (amount or size) or quality. The motion starts when the item is first picked up or viewed and ends when it is either released or used in assembly.

Pre-Position: (a bowling pin being placed into proper position)

This is the motion of replacing an item in the proper orientation for its next Use. In the example of the pen being in a holder on the table, the act of replacing the pen, in the proper Position for its next use would be pre-position. Like position, it can be performed during transport loaded. Frank Gilbreth's favorite example was when a pool shot is planned so that the cue ball ends up in a good position for the next shot.

Release Load: (a hand with an object poised to drop)

This motion involves releasing the object when it reaches its destination. The actual time taken will be fractions of a second and would vary with such things as if it were being Pre-Positioned, or if the release was merely down a hole, into a gravity chute. Caution must be taken in solely working towards short release times. For example, it may be quicker to drop the part into a bin, but what about the next station/operation?

Unavoidable Delay: (a man bumping his nose unintentionally)

This Therblig is measured from the point where a hand is inactive to the point where it becomes active again, with another Therblig. These delays were defined by Gilbreth, as being out of the control of the particular worker being studied. They could involve a lack of raw materials being available or repair of a tool, etc. While these delays might be dealt with by the overall factory/business system, they are not considered the responsibility of the individual operator.

Avoidable Delay: (a worker intentionally lying down on the job)

This counterpart to unavoidable delay involves inactive time the worker encounters over which he/she has control. For example, if the worker is required to do inspections of his/her tools and report problems, and the result is that a tool that breaks in the middle of the shift, the worker is responsible for the delay. Avoidable delays can also occur with an individual hand or a body part, which remains idle while another is working harder than needed.

Plan: (a worker with fingers on head, thinking)

This Therblig is a mental function, which may occur before Assemble (deciding which part goes next) or prior to Inspection, noting which flaws to look for. The extent of the use of Plan varies greatly with the type of job performed. However, in routine jobs, the time spent in the Plan Therblig should be kept to a minimum through arrangement of parts and tools.

Rest to overcome Fatigue: (a person resting in a seated position)

This Therblig is actually a lack of motion and is only found where the rest is prescribed by the job or taken by the worker. In the Gilbreths' scheme of Fatigue Reduction, after you had eliminated all unnecessary motions and made necessary ones as least fatiguing as possible, there would still be the need to rest.

3.4 Motion Study

Motion study is a technique of analysing the body motions employed in doing a task in order to eliminate or reduce ineffective movements and facilitates effective movements. By using motion study and the principles of motion economy the task is redesigned to be more effective, and less time consuming.

The Gilbreths pioneered the study of manual motions and developed basic laws of motion economy that are still relevant today. They were also responsible for the development of detailed motion picture studies, termed as Micro Motion Studies, which are extremely useful for analysing highly repetitive manual operations. With the improvement in technology, of course video camera has replaced the traditional motion picture film camera.

In a broad sense, motion study encompasses micro motion study and both have the same objective: job simplification, so that it is less fatiguing and less time consuming. While motion study involves a simple visual analysis, micro-motion study uses more expensive equipment. The two types of studies may be compared to viewing a task under a magnifying glass versus viewing the same under a microscope. The added detail revealed by the microscope may be needed in exceptional cases when even an improvement of a minute in motions matters, i.e.,in extremely short repetitive tasks.

Taking the cine films @ 16 to 20 frames per second with motion picture camera, developing the film and analysing the film for micro-motion study had always been considered a costly affair. To save on the cost of developing the film and the cost of film itself, a technique was used in which camera took only 5 to 10 frames per minute. This saved the time of film analysis too. In applications where infrequent shots of camera could provide almost same information, the technique proved fruitful and acquired the name Memo Motion Study.

Traditionally, the data from micro-motion studies are recorded on a Simultaneous Motion (SIMO) Chart, while that from motion studies are recorded on a Right Hand - Left Hand Process Chart.

3.5 Cycle Graph and Chronocycle Graph

These are the techniques of analysing the paths of motion practiced by an operator and were originally developed by Gilbreths. These techniques are used for recording the path of movements too. They are photographic methods and are accurate and detailed. They are used to trace especially those movements which are too fast for human eye to record.

Cycle Graph: A cyclegraph system records the path of movements of body part of an operator while he/she is performing the operation. To make a cycle graph, a small electric bulb is attached to the finger, hand, or any other part of the body, whose motion is to be recorded. By using still photography, the path of light of bulb as it moves through space for one complete cycle is photographed. The working area is generally kept less illuminated while photograph is being taken. More than one camera may be used in different planes to get more details. After the film is developed, the resulting picture (cycle graph) shows a permanent record of the motion pattern employed in the form of a closed loop of white continuous line with the working area in the background. A cycle graph does not indicate the direction or speed of motion. Figure 3.9 (a) & (b) show cycle and Chronocycle Graph, respectively.

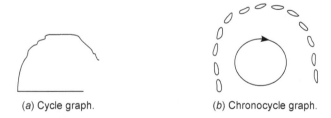

(a) Cycle graph.   (b) Chronocycle graph.

Fig. 3.9 Cycle graph(a) and Chronocyclegraph(b)

It can be used for

- Improving the motion pattern, and
- Training purposes where two cycle graphs may be shown with one indicating a better motion pattern than the other.

Chronocycle Graph: Cyclegraph neither tells the direction nor the speed of the movements while performing the operation. The chronocycle graph is similar to the cycle graph, but the power supply to the bulb is interrupted regularly by using an electric circuit which varies from 10 to 30 sec.). The bulb is thus made to flash. Rest of the procedure for taking photograph remains the same. The resulting picture (chronocycle graph), instead of showing continuous line of motion pattern, shows short dashes of line spaced in proportion to the speed of the body member photographed on photographic plate. Wide spacing would represent fast moves while close spacing would represent slow moves. The jumbling of dots at one point would indicate fumbling or hesitation of the body member. A chronocycle graph can thus be used to study the motion pattern as well as to compute velocity, acceleration and retardation experienced by the body member at different locations. Figures show a cycle graph and a chronocycle graph. The world of sports has extensively used this analysis tool, updated to video, for the purpose of training in the development of form and skill.

3.5.1 Equipment Used

- A chronographic unit, which is a device (formerly battery operated) to light the bulbs and to enable the flashing to vary between 10 and 25 flashes per second;
- Camera and exposure meter - a camera capable of double exposure will enable a normal picture of the scene to be superimposed over the chronocycle graph;
- Supplementary lighting for use when taking the instantaneous exposure to obtain the superimposed picture.

3.5.2 Procedure of Chronocycle Graph

1. Set the frequency of flashes depending on the type of job;
2. Attach bulbs to hands;
3. Decide the exposure time;
4. Take a time exposure for the period of the job cycle, whilst the operator performs the job the location would be a darkened room in former times;
5. Take a second, instantaneous exposure on the same film, if it is required to superimpose the scene on the chronocycle graph.

3.5.3 Uses of Chronocycle Graph

1. While developing a better workplace Chronocycle graphs reveal obstructions and bad locations;
2. Analysis of a complex movement;
3. An aid to training;
4. Comparison of two methods;
5. Publicity and advertising;
6. Design of new equipment.

3.6 Principles of Motion Economy

A set of rules were designed by Gilbreth in order to develop better methods. A better method of doing a job is one which consumes minimum of time and energy in performing body parts motions in order to complete the task, and it is possible only by economising the use of motions. The rules of human motion as presented by Gilbrethwere rearranged and amplified by Lowry, Maynard and Barnes others. Its various rules are as follows:

- Those related to the use of the human body.
- Those related to the workplace arrangement, and
- Those related to the design of tools and equipment.

3.6.1 Principles Related to the use of Human Body

(*i*) Both hands should begin and end their basic divisions of activity simultaneously and should not remain idle at the same instant, except during the rest periods.

(*ii*) The hand motions should be made symmetrically and simultaneously away from and toward the centre of the body.

(*iii*) Momentum should be employed to assist the worker wherever possible.

(*iv*) Continuous curved motions should be preferred to straight line motions involving sudden and sharp changes in the direction.

(*v*) The least number of basic divisions should be employed and these should be confined to the lowest practicable classifications. These classifications, summarised in ascending order of time and fatigue expended in their performance, are:

(*a*) Finger motions

(*b*) Finger and wrist motions.

(*c*) Finger, wrist, and lower arm motions.

(*d*) Finger, wrist, lower arm, and upper arm motions.

(*e*) Finger, wrist, lower arm, upper arm motions and body motions.

(*vi*) Work that can be done by feet should be arranged so that it is done together with work being done by the hands. It should be recognised, however, that it is difficult to move the hand and foot simultaneously.

(*vii*) The middle finger and the thumb should be used for handling heavy loads over extended periods as these are the strongest working fingers. The index finger, fourth finger, and little finger are capable of handling only light loads for short durations.

(*viii*) The feet should not be employed for operating pedals when the operator is in standing position.

(*ix*) Twisting motions should be performed with the elbows bent.

(*x*) To grip tools, the segment of the fingers closed to the palm of the hand should be used.

3.6.2 Principles Related to the Arrangement and Conditions of Workplace

(*i*) Fixed locations should be provided for all tools and materials so as to permit the best sequence and eliminate search and select.

(*ii*) Gravity bins and drop delivery should be used to reduce reach and move times. Use may be made of ejectors for removing finished parts.

(*iii*) All materials and tools should be located within the normal working area in both the vertical and horizontal plane, and as close to the point of use as possible.

(*iv*) Worktable height should permit the operator to alternately sit and stand while working.

(*v*) Glare-free adequate illumination, proper ventilation and proper temperature should be provided.

(*vi*) Dials and other indicators should be patterned such that maximum information can be obtained in minimum of time and error.

3.6.3 Principles Related to the Design of Tools and Equipment

(*i*) Use colour, shape or size coding to maximise speed and minimise error in finding controls.

(*ii*) Use simple on/off, either/or indicators whenever possible. If simple on/off indicator is not sufficient, use qualitative type indicator, and use quantitative type indicator only when absolutely essential.

(*iii*) All levers, handles, wheels and other control devices should be readily accessible to the operator and should be designed so as to give the best possible mechanical advantage and utilise the strongest available muscle group. Their direction of motion should conform to stereotyped reactions.

(*iv*) Use quick acting fixture to hold the part or material upon which the work is being performed.

(*iv*) Use stop guides to reduce the control necessary in positioning motions.

(*v*) Operating, set-up and emergency controls should be grouped according to the function.

Summary

This chapter deals with the definitions, objectives and step by step procedure of Method Study. It also discusses various recordings techniques, such as Operation Process Chart, Flow Process Chart, Multiple Activity Chart, Two Handed Process Chart, Travel Chart and String Diagram along with their applications. Apart from this, it also highlights the micro-motion study, applications of SIMO Chart and Principles of Motion Economy.

Work Measurement

4.1 Definition and Objectives

Work measurement refers to the estimation of standard time taken, implying the time allotted for completing a piece of job by using the prescribed method. Standard time can be defined as the time taken by an average experienced worker for the job, with provisions for delays beyond the worker's control.

There are several techniques used to estimate the standard time in industry. These include time-study, work-sampling, standard data, and predetermined motion-time system.

Work Measurement is the application of techniques designed to establish the time for an average worker to carry out a specified manufacturing task, at a defined level of performance. It is concerned with the length of time taken to complete a specific task assigned to a worker Work measurement means to estimate of standard time taken to complete a given piece of job by using the prescribed method.

Work measurement may be defined as "the art of observing and recording the time required to do each detailed element of an industrial activity/operation."

The term industrial activity includes mental, manual and mechanical operations, where:

(*i*) Mental time includes time taken by the operator for thinking over some alternative operations.

(*ii*) Manual time consists of three types of operations, *i.e.,* one related to the handling of materials, second to the handling of tools, and third to the handling of machines.

Objectives

Work measurement helps to uncover non-standardisation that exists in the workplace and non-value adding activities and waste. A work has to be measured for the following reasons:

- To discover and eliminate lost or unproductive time
- To establish standard time for performance measurement
- To measure performance against realistic expectations
- To set operating goals and objectives

Applications

Estimation of standard time for operations is useful for several applications in industry, like

- Estimating material, machinery, and equipment requirements
- Estimating production cost per unit as an input to
- Preparation of budgets
- Determination of selling price
- Decision for making or buying
- Estimating manpower requirements
- Estimating delivery schedules and planning the work
- Balancing the distribution of the work of operators working in a group
- Estimating the performance of workers and using it as the basis for payment of incentive to those direct and indirect or labour who show greater productivity

4.2 Time-Study Procedure

The procedure for time study can best be described step wise, which are self-explanatory.

Step 1: Define the objective of the study. This involves using the statement of the result, the desired precision, and the required level of confidence in the estimated time standards.

Step 2: Verify that the standard method and conditions exist for the operation, and the operator is properly trained. If the need is felt for method study or further training of the operator, it should be completed before starting the time study.

Step 3: Select the operator to be studied, if there are more than one operators doing the same task.

Step 4: Record information about the standard method, operation, operator, product, equipment, and conditions on the Time study observation sheet.

Step 5: Divide the operation into reasonably small elements, and record them on the Time Study observation sheet.

Step 6: Time the operator for each of the elements. Record the data for a few numbers of cycles on the Time Study observation sheet. Use the data to estimate the total number of observations taken.

Step 7: Collect and record the data of required number of cycles by timing and rating the operator.

Step 8: Calculate the watch time taken by the representive for each element of operation. Multiply it by the rating factor to get the normal time.

Normal time = Observed time x Rating factor

Calculate the normal time for the whole operation by adding the normal time of its various elements.

Step 9: Determine allowances for fatigue and various delays.

Step 10: Determine standard time of operation.

Standard time = Normal time + allowances

4.3 Work-Measurement Techniques

For the purpose of work measurement, work can be regarded as:

1. **Repetitive work:** The type of work in which the main operation or group of operations repeat continuously during the time spent at the job. These apply to work cycles of extremely short duration.

2. **Non-repetitive work:** It includes those kinds of maintenance and construction work where the work cycle itself is hardly ever repeated identically.

Techniques of Work Measurement

1. Stopwatch Time Study
2. Work-Sampling
3. Analytical estimating
4. Synthesis

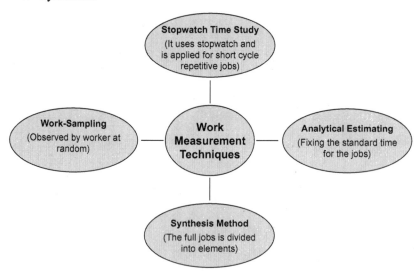

Fig. 4.1 Work-Measurement Technique

Time study and work-sampling involve direct observation, whereas the remaining are data-based and analytical in nature.

1. **Stopwatch Time Study:** It is a work measurement technique for recording the time and rates of working for the elements of a specified job carried out under specified conditions for analysing the data so as to determine the time necessary for carrying out the job at the defined level of performance. In other words measuring the time through stopwatch is called time study.

2. **Synthetic data:** It is a work-measurement technique for building up the time for a job, or pans of the job, at a defined level of performance, by totalling element time obtained previously from time studies on other jobs containing the elements concerned or from synthetic data.

3. **Work-sampling:** This is a technique in which a large number of observations are made over a period of time on one or a group of machines, processes or workers. Each observation records what is happening at that instant, and the percentage of observations is recorded for a particular activity, which or its delay, is a measure of the percentage of time during which the activities delay had occured.

4. **Predetermined motion time study (PMTS):** A work-measurement technique whereby time established for basic human motions (classified according to the nature of motions and conditions under which they are made) is used to build up the time for a job on the defined level of performance. The most commonly used PMTS is known as Methods Time Measurement (MTM).

5. **Analytical estimating:** This is a work-measurement technique, a development of estimating whereby the time required to carry out elements of a job on a defined level of performance is estimated partly from knowledge and practical experience of the elements concerned and partly from synthetic data.

The work-measurement techniques and their applications are shown in the following table.

Table 4.1: Work-Measurement techniques and their applications

Techniques	Applications	Unit of measurement
1. Time Study	Short cycle repetitive jobs widely used for direct work.	Centiminute (0.01 min)
2. Synthetic Data	Short cycle repetitive jobs.	Centiminutes
3. Working-Sampling	Long cycle jobs/heterogeneous operations.	Minutes
4. MTM	Manual operations confined to one work centre.	TMU (1 TMU = 0.006 min)
5. Analytical Estimation	Short cycle non-repetitive job.	Minutes

4.3.1 Stopwatch Time Study

Time study was developed by F.W. Taylor, and since a stopwatch is used for making time observations it is called stopwatch time study. It is a unique technique of work measurement as it involves direct observation of work while it is performed. Time study can be defined as follows:

Time study is a work measurement to facilitate the management to establish the time standards for a particular job or its elements. The established standards can be used as a benchmark to compare the results achieved in future, and exercise the necessary control. Direct time study is mainly useful for repetitive work, i.e., any job which is subsequently going to be repeated under the given circumstances and method established in method study.

The most important feature of time study is the approach of data recording, which surely affects the accuracy of the results obtained. The precision results directly depend upon the number of observations taken, i.e., more the number of observations the more accurate will be the results obtained from time study. Therefore, the accuracy of time study increases with the number of observations conducted to make the study.

Stopwatch types

A stopwatch is one of the principle timing devices employed for measuring the time taken by the operator to complete a job. It is also an accurate time-measuring equipment that can run continuously for one hour or half an hour normally, and records time by its small hand. One revolution of the big hand of the stopwatch which records one minute, and the scale covering one minute may be calibrated in the intervals of $1/100^{th}$ of a minute or $1/300^{th}$ of a minute. The different types of stopwatches are as follows:

(*i*) Non-fly back

(*ii*) Fly back

(*iii*) Split hand or split-second type

Non-fly back stopwatch

Non-fly back stopwatch is exhibited in figures 4.1. It is mainly preferred for measuring continuous timings. With first pressing of the winding knob the watch starts and the long hand begins to move. If winding knob is pressed again, the long hand pauses and with the third pressing, hands come to their initial position. In the case of measuring timings in two parts, where the second part happens right after the first, the non-fly back system does not work well, because it involves stopping the watch at the end of the first part, pressing the knob to bring the hands back to zero, and again pressing the knob to start the hands. This consumes quite some time and leaves less margins for timing the second element. Hence the second element cannot be timed accurately. In such a case the use of fly back or split-hand type of stopwatch is required.

Fly back stopwatch

Non-fly back stopwatch is exhibited in figure 4.2. In fly back system, slide is used to start and stop the watch. The hands come to zero by pressing the winding knob. However, they do not stop and begin straightway moving forward again. The slide is employed to stop the hands at any point. This stopwatch is preferred for taking fly back timings or continuous timing observations,hence it can easily give precise reading.

Split-hand type of watch

This stopwatch provides more accuracy in reading, particularly when two parts are to be timed, and the second right away follows the first one. As one part ceases, pressing the winding knob makes one hand to stop, whereas the other keeps moving. Once the reading has been taken, a second pressing on the knob stops and restarts the hand, and then the two hands go along.

4.3.2 Work-Sampling

Work-Sampling (also sometimes called ratio-delay study) is a technique of getting facts about the utilisation of machines or human beings through a large number of instantaneous observations taken at random intervals. The ratio of observations of a given activity to the total observations approximates the percentage of time that the process is in which state of activity. For example, if 500 instantaneous observations taken at random intervals over a few weeks show that a lathe operator was doing productive work in 365 observations, and in the remaining 135 observations he was found 'idle' for miscellaneous reasons, then it can be reliably taken that the operator remains idle (135/500) x 100 = 27 % of the time. Obviously, the accuracy of the result depends on the number of observations. However, in most applications there is usually a limit beyond which a greater accuracy of data is not economically viable.

Use of Work-Sampling for Standard Time Determination

Work-sampling can be very useful for establishing time standards on both direct and indirect labor jobs. The procedure for conducting work-sampling study for determining the standard time of a job can be described stepwise as follows.

Step 1: Define the problem.

- Describe the job for which the standard time is to be determined.

- Unambiguously state and discriminate between the two classes of activities of operator on the job: what are the activities of job that would entitle him to be in ‹working» state.

This would imply that when operator will be found engaged in any activity other than the those which come in the category of "working", he/she would be categorized in "Not working" state.

Step 2: Design the sampling plan.

- Estimate satisfactory number of observations to be made.

- Decide on the period of study, like two days, one week, etc.

- Preparea detailed plan for taking the observations.

   This will include observation schedule, exact method of observing, design of observation sheet, route to be followed, particular person to be observed at the decided time, etc.

Step 3: Contact the persons concerned and take them in confidence about conducting the study.

Step 4: Make observations at predecided though random times about the working / not working state of the operator. When operator is in working state, determine his performance ratings. Record both on the observation sheet.

Step 5: Obtain and record other information too. This includes operator›s starting time and quitting time of the day, and total number of parts of acceptable quality productivity during the day.

Step 6: Calculate the standard time per piece.

We will now briefly discuss some important issues involved in the procedure.

Number of Observations

As we know, results of a study based on larger number of observations would be more accurate, but taking more and more observations consumes time, and is costly thus. A cost-benefit trade-off has thus to be struck. In practice, the following methods are used to estimate the number of observations to be made.

(*i*) **Based on judgment.** The study person can decide the necessary number of observations based on his/her judgment. The correctness of the number may be doubtful, but estimate is often quick and in most cases adequate.

(ii) **Using cumulative plot of results.** As the study progresses the results of the proportion of time devoted to the given state or activity, i.e., Pi from the cumulative number of observations are plotted at the end of each shift or day. A typical plot is shown in Figure. Since the accuracy of the result improves with increasing number of observations, the study can be continued until the cumulative Pi appears to stabilise, and the collection of further data would seem to have a negligible effect on the value of Pi.

(iii) **Use of statistics.** In this method, by considering the importance of the decision that would be based on the results of study, a maximum tolerable sampling error in terms of confidence level and desired accuracy in the results is specified. A pilot study is then made in which a few observations are taken to obtain a preliminary estimate of Pi. The numbers of observations N necessary are then calculated using the following expression.

The number of observations estimated from the above relation using a value of Pi obtained from a preliminary study would be only a first estimate. In actual practice, as the work-sampling study proceeds, say at the end of each day, a new calculation should be made by using increasingly reliable value of Pi obtained from the cumulative number of observations made.

Determination of Observation Schedule

The number of instantaneous observations to be made each day mainly depends upon the nature of operation. For example, for non-repetitive operations, or for operations in which some elements occur infrequently, it is advisable to take observations more frequently so that the chances of obtaining all the facts improve. It also depends on the availability of time with the person making the study. In general, about 50 observations per day is a good figure. The actual random schedule of the observations is prepared by using a random-number table, or any other technique.

Design of Observation Sheet

A sample observation sheet for recording the data with respect to whether at the predecided time the specified worker on the job is in 'working' state or 'non-working' states is shown in Figure. It contains the relevant information about the job, and the operators on the job, etc. At the end of each day, calculation can be done to estimate the percent of time workers on the job (on an average) spend on activities, which are considered as part of their job.

Conducting Work-Sampling Study

At predecided times of study, the study person appears at the work site and observes the specific worker (already randomly decided) to find out what is he/she doing. If he/she is doing an activity which is a part of the job, he/she is ticked under the column 'Working', and his/her performance rating is estimated and recorded. If he/she is found engaged in an activity which is not a part of the job, he/she is ticked under the column 'Not Working'. At the end of day, the number of ticks in 'Working' column is totalled and average performance rating is determined.

The observed time (OT) for a given job is estimated as

$$S.P_1 = x\sqrt{\frac{P_1(1-P_1)}{N}}$$

Where   S = desired relative accuracy

P_1 = estimate of proportion of time devoted to activity, expressed as a decimal, *e.g.,* 5% = 0.05

x = a factor depending on the confidence level.

x = 1, 2, 3 for confidence levels of 68%, 95% and 99% respectively.

N = total number of observations needed.

The normal time (NT) is found by multiplying the observed time by the average performing index (rating factor).

$$NT = OT \times \left(\overline{R}/100\right)$$

Where $= \overline{R}$ is average rating factor to be determined as $\dfrac{\left(\Sigma^R\right)}{\eta_1}$,Figure

The standard time is determined by adding allowances to the normal time.

Advantages and Disadvantages of Work-Sampling in Comparison with Time Study

Advantages

1. **Economical**

   - Many operators or activities which are difficult or uneconomical to measure by time study can readily be measured by work-sampling.

   - Two or more studies can be simultaneously made of several operators or machines by a single study person. Ordinarily a work-study engineer can study only one operator at a time when continuous time study is made.

   - It usually requires fewer man-hours to make a work-sampling study than to make a continuous time study. The cost may also be about a third of the cost of a continuous time study.

   - No stopwatch or other time-measuring device is needed for work-sampling studies.

   - It usually requires less time to calculate the results of a work-sampling study. Mark sensing cards may be used which can be fed directly to the computing machines to obtain the results instantaneously.

2. **Flexible**

   - A work-sampling study may be interrupted at any time without affecting the results.

   - Operators are not closely watched for a long period of time. This decreases the chance of getting erroneous results for when a worker is observed continuously for a long period, it is probable that he will not follow his usual routine exactly during that period.

3. Less Erroneous

- Observations may be taken over a period of days or weeks. This decreases the chances of day-to-day or week-to-week variations that may affect the results.

4. Operators Like It

- Work-sampling studies are preferred to continuous time study by the operators being studied. Some people do not like to be observed continuously for long periods of time.

5. Observers Like It

- Work-sampling studies are less fatiguing and less tedious to make on the part of time-study engineer as well.

Disadvantages

- Work-sampling is not economical for the study of a single operator or operation, or a single machine. Also, work-sampling study may be uneconomical for studying operators or machines located over wide areas.

- Work-sampling study does not provide elemental time data.

- The operator may change his work pattern when he sees the study person. For instance, he may try to look productive and make the results of study erroneous.

- No record is usually made of the method being used by the operator. Therefore, a new study has to be made when a method change occurs in any element of operation.

- Compared to stopwatch time study, the statistical approach of work-sampling study is difficult to understand by workers.

4.4 Job Selection for Work Measurement

Time Study is conducted on a job

- which has not been previously time-studied.
- for which method change has taken place recently.
- for which worker(s) might have complained as having tight time standards.

4.5 Equipments and Forms used for Time-Study

Time-Study Equipment

The following equipment is needed for time-study work.

- (*a*) Timing device
- (*b*) Time-study observation sheet
- (*c*) Time-study observation board
- (*c*) Other equipments

- (*a*) **Timing Device.** The stopwatch is the most widely used timing device for time study, although electronic timer is also sometimes used. The two perform the same function with the difference that electronic timer can measure time to the second or third decimal of a second and can keep a large volume of time data in memory.

- (*b*) **Time-Study Observation Sheet.** It is a printed form with spaces provided for noting down necessary information about the operation being studied, like the name of operation, drawing number, and the name of the worker, name of time-study person, and the date and place of study. Spaces are provided in the form for writing detailed descriptions of the process (element wise), recorded time or stopwatch readings for each element of the process, performance rating(s) of operator, and computation.

- (*c*) **Time-Study Board.** It is a light-weight board used for holding the observation sheet and stopwatch in position. In size it is slightly larger than the observation sheet used. Generally, the watch is mounted at the center of the top edge near the upper right-hand corner of the board. The board has a clamp to hold the observation sheet. During the time study, the board is held against the body and the upper left arm by the time-study person in such a way that the watch could be operated by the thumb/ index finger of the left hand. Watch readings are recorded on the observation sheet by the right hand.

- (*d*) **Other Equipment.** This includes pencil, eraser, devices like tachometer for checking the speed, etc.

4.6 Performance Rating

During the time-study, the engineer carefully observes the performance of the operator. This performance seldom conforms to the exact definition of normal or the set standard. Therefore, it becomes necessary to apply some 'adjustment' to the mean observed time to arrive at the time that the normal operator would have taken to do that job, when working at an average pace. This 'adjustment' is called Performance Rating.

Determination of performance rating is an important step in the work-measurement procedure. It is based entirely on the experience, training, and judgment of the work-study engineer. Since it is the most subjective step, therefore is subject to criticism undoubtedly.

Performance Rating can be defined as the procedure in which the time study engineer compares the performance of operator(s) under observation to the Normal Performance and determines a factor called Rating Factor.

$$\text{Rating Factor} = \frac{\text{Observed Performance}}{\text{Normal Performance}}$$

System of Rating

There are several systems of rating the performance of a operator on a job.

(*a*) Pace Rating

(*b*) Westing house System of Rating

(*c*) Objective Rating

(*d*) Synthetic Rating

A brief description of each rating method follows:

(*a*) **Pace Rating**

Under this system, operator's performance is evaluated by considering his rate of accomplishment of the work. The study person measures the effectiveness of the operator against the concept of normal performance, then he/she assigns a percentage to indicate the ratio of the observed performance to normal or standard performance.

In this method, which is also called the speed rating method, the time-study person judges the operators speed of movements, *i.e.,* the rate at which he is applying himself/herself, or in other words "how fast" the operator is performing the motions involved.

(b) Westing House System of Rating

This method considers four factors in evaluating the performance of an operator: skill, effort, conditions, and consistency.

Skill may be defined as the proficiency of an individual in following the given method. It is demonstrated by the coordination of mind and hands. A person's skill in a given operation increases with his experience on the job, because increased familiarity with work brings speed, smoothness of motions, and freedom from hesitations.

The Westing house system lists six classes of each factor. For instance the classes of skill are categorised as poor, fair, average, good, excellent and super skill in order of excellence, as given in the Table. The time-study person evaluates the skill displayed by the operator by adding two degrees extra as the performance then he/she puts it in one of the six classes after deciding the degree of that class, higher or lower, *i.e.,* 1 or 2. The equivalent % value of each class of skill is provided in the Table, the rating is translated into its equivalent percentage value, which ranges from +15 % (for super skill of higher degree) to -22 % (for poor skill of lower degree).

In a similar fashion, the ratings for effort, conditions, and consistency are given in the Table for each of the factors. By algebraically combining the ratings with respect to each of the four factors, the final performance-rating factor is estimated.

(c) Objective Rating

In this system, speed of movements and job difficulty are rated separately and the two estimates are combined into a single value. Rating of speed or pace is done as discussed earlier, and the rating of job difficulty is done by selecting adjustment factors corresponding to the characteristics of the operation with respect to (*i*) amount of body used, (*ii*) foot pedals, (*iii*) bi-manualness, (*iv*) eye-hand coordination, (*v*) handling requirements, and (vi) weight handled or resistance encountered. Mundel and Danner have given a Table of % values (adjustment factors) for the effects of various difficulties in the operation performed.

For an operation under study, a numerical value for each of the six factors is assigned, and the algebraic sum of the numerical values called job difficulty adjustment factor is estimated.

The rating factor R can be expressed as

$$R = P \times D$$

Where: P = Pace rating factor, and

                    D = Job difficulty adjustment factor.

(*d*) Synthetic Rating

This method of rating has two main advantages over other methods. Firstly it does not rely on the judgment of time-study person, and secondly it gives consistent results.

The time study is made as usual. Some manually controlled elements of the work cycle are selected. Using a PMT system (Predetermined motion time system), the times for these selected elements are determined. These a determined elements are then compared with the actual observed times, and then the performance factor is estimated for each of the selected elements.

Performance or Rating Factor,

$$R = P / A$$

Where   P = Predetermined motion time of the element, and

            A = Average actual observed time of the element.

The overall rating factor is the mean of rating factors determined for the selected elements. This is applied uniformly to all the manually controlled elements of the work cycle.

4.7 Determination of Normal Time and Standard Time Allowances

The readings of any time study are taken over a relatively short period of time. The normal time arrived at, therefore, does not include unavoidable delay and other legitimate lost time: for example, in waiting for materials, tools or equipment; periodic inspection of parts; interruptions due to legitimate personal needs, etc. It is necessary and important that the time-study person applies some adjustment, or allowances, to compensate for such losses so that a fair time standard is established for the given job.

Allowances are generally applied as some percentage of total cycle time, some percent is given separately for machine time, and some percent for manual effort time. However, no allowances are given for interruptions which may occur due to the factors which are within the operator's control or which are avoidable.

There are a number of factors responsible for the same. However some of the most important among them are as follows;

1. **Factors related to the individual worker:** In case every worker is to be considered individually in a particular working area. A thin, active and alert worker at the peak of his/her physical health condition will require a smaller allowance to recover from fatigue than an obese, clumsy and weak worker. Likewise, each worker has his/her own learning curve which can affect the way in which he/she performs his/her job. There is also some reason to believe that there may be cultural/ regional/racial variations in response to the level of fatigue experienced by workers, mainly when engaged on heavy physical workload.

2. **Factors related to the nature of the work:** Most of the methods established for the calculations of allowances may be applicable to light and medium jobs, but at the same time these may be found inadequate for very heavy and strenuous activities, such as lifting and carrying of molten metal in a crucible from furnaces to the moulds in casting firms.

3. Factors related to environment: The development of a standardized method to meet allowances is quite difficult for every work situation affected by various workplace factors, such as level of heat stress, intensity of light, level of noise, dust, and fume exposure. Many of the workplace environment factors are season-dependant. Now the reader may be clear in his/her vision that the ILO has not adopted any allowance calculating standard.

Most companies allow the following allowances to their employees:

- Constant allowances (for personal needs and basic fatigue)
- Delay Allowance (for unavoidable delays)
- Fatigue Allowance (for job-dependent fatigue)
- Personal Allowance
- Special Allowance

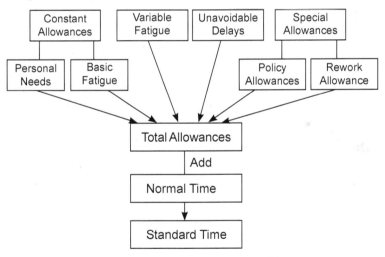

Fig. 4.2 Determination of Standard Time

Delay Allowance

This time allowance is given to the operator for the numerous unavoidable delays and interruptions that he/she experiences every day during the course of his/her work. These interruptions include interference from the supervisors, inspectors, planners, expediters, fellow workers, production personnel and others. This allowance also covers interruptions occurring due to material irregularities, difficulty in maintaining specifications and tolerance, and interference delays where the operator has to attend to more than one machine.

Fatigue Allowance

This allowance can be divided into two parts: (*i*) basic fatigue allowance and (*ii*) variable fatigue allowance. The basic fatigue allowance is given to the operator to compensate for the energy expended for carrying out the work and to alleviate monotony. For an operator who is doing light work while seated, under good working conditions, and under normal demands on the sensory or motor system, a 4% of normal time is considered adequate. This can be treated as a constant allowance.

The magnitude of variable fatigue allowance given to the operator depends upon the severity of conditions, which cause extra (more than normal) fatigue to him. As we know, fatigue is not homogeneous. It ranges from strictly physical to purely psychological and includes combinations of the two. On some people it has a marked effect while on others, it has apparently little or no effect. Whatever may be the kind of fatigue-physical or mental, the result is same—it reduces the work output of operator. The major factors that cause more than just the basic fatigue includes severe working conditions, especially with respect to noise, illumination, heat and humidity; the nature of work, especially with respect to posture, muscular exertion and tediousness, and the like.

It is true that in modern industry, heavy manual work, is reducing day by day, hence the muscular fatigue, but mechanisation is promoting other fatigue components, like monotony and mental stress. Because fatigue in totality cannot be eliminated, proper allowance has to be given for adverse working conditions and repetitiveness of the work.

Personal Allowance

This is allowed to compensate for the time spent by the worker in meeting the physical needs, for instance, a periodic break in the production routine. The amount of personal time required by the operator varies with the individual more than with the kind of work, though it is seen that workers need more personal time when the work is heavy and done under unfavourable conditions. The amount of this allowance can be determined by making all-day time study or work-sampling. Mostly a 5 % allowance for personal time (nearly 24 minutes in 8 hours) is considered appropriate.

Special Allowances

These allowances are given under certain special circumstances. Some of these allowances and the conditions under which they are given are:

(*a*) **Policy Allowance:** Some companies, as a policy, give an allowance to provide a satisfactory level of earnings for a specified level of performance under exceptional circumstance. This may be allowed to new employees, handicap employees, workers on night shift, etc. The value of the allowance is typically decided by management.

(*b*) **Small Lot Allowance:** This allowance is given when the actual production period is too short to allow the worker to come out of the initial learning period. When an operator completes several small-lot jobs on different setups during the day, an allowance as high as 15 percent may be given to allow the operator to make normal earnings.

(*c*) **Training Allowance:** This allowance is provided when work is done by a trainee to allow him to make reasonable earnings. It may be a sliding allowance, which progressively decreases to zero over a certain length of time. If the effect of learning on the job is known, the rate of decrease of the training allowance can be set accordingly.

(*d*) **Rework Allowance:** This allowance is provided on certain operations when it is already known that some percent of parts made are spoiled due to factors beyond the operator's control. The time in which these spoiled parts may be reworked is converted into allowance.

Different organisations have decided upon the amount of allowances to be given to different operators by taking help from the specialists / consultants in the field and through negotiations between the management and the trade unions. ILO has given its recommendations about the magnitude of various allowances, as shown in the following Table.

Basic time

It is the time required to perform a task by a normal operator working at a standard pace(rate) with no allowances for personal delays, unavoidable delays, or fatigue.

$$\text{Basic Time (Normal Time)} = \text{Observed time} \times \frac{\text{Observed Rating}}{\text{Standard Rating (100)}}$$

$$\text{BT (or NT)} = \text{OT} \times \frac{\text{OR}}{\text{SR(100)}} \quad i.e., \ \text{OT} \times \text{OR} = \text{BT} \times \text{SR}$$

Basic (Normal) Time can be represented as

$$= (\text{Observed Time}) \times \left(\frac{\text{Rating factor in \%}}{100} \right)$$

Rating factor (RF) is in percentage like if observed rating is 90 per cent, RF = 0.9, if rating is 110 per cent then R.F = 1.10

Where, BT : Basic time

NT : Normal time

OT : the observed time with the stopwatch

OR : Observed rating

SR : Standard rating

Allowed time

It is the time obtained by adding the percentage of rest allowance, process allowance, and the special allowance, in basic or normal time. It can be calculated as:

$$\text{Allowed time} = (\text{Normal Time}) \times \left(1 + \frac{\text{Allowances in \%}}{100} \right)$$

Standard time

When all the rest allowances are considered alone, it is called sustained performance time, the calculation of the same is illustrated in the example give ahead. It is obtained by adding the policy allowance to the allowed time.

$$\text{Standard time (ST)} = (\text{Allowedtime}) \times \left(1 + \frac{\text{Policy Allowances in \%}}{100}\right)$$

Thus the standard time is ultimately determined from the 'observed time' noted from the stopwatch time study or alternately it can be represented as follows:

$$\text{Standard time (ST)} = (\text{Normal time} \times \text{Allowance factor})$$

Where,   Allowance factor $= \left(\dfrac{1}{1 - \text{Allowances in \%}}\right)$

Numerical 1: Calculate the Standard time for the data given in the table 2, using the above mentioned method.

Table 4.2 Data Analysis

Element No.	1	2	3	4
Cycle No.	Stopwatch Reading in seconds			
1	20	30	45	65
2	100	150	175	210
3	220	240	245	250
4	265	270	275	285
5	300	320	330	340
Average rating	181	202	214	230

Solution: The initial stopwatch reading is taken as zero, the observed time for each element is determined from the continuous readings taken at the end of each element by subtracting the initial from the final. Therefore observed timings for each element is shown in table.

Work Study and Ergonomics

Table 4.3 Determination of Normal Time

Element No.	1	2	3	4
Cycle No.		Stopwatch Reading in seconds		
1	20	10	15	20
2	35	50	25	35
3	10	20	5	5
4	15	5	5	10
5	15	20	10	10
Average	19	21	12	16
Average rating	181	202	214	230
Normal Time	(19*181)/100 =34.39	(21*202)/100 =42.42	(12*214)/100 =25.68	(16*230)/100 =36.80

Therefore normal time for the operation

$$= 34.39+42.42+25.68+36.80 = 139.29 \text{ sec}$$

The given fatigue allowance = 10%

Therefore sustained performance time

$$= (\text{Normal Time})*110/100 = (139.29*110)/100$$

$$= 153.22 \text{ sec}$$

Total Fatigue Process and contingency allowances

$$= 10 + 5 + 5 = 20\%$$

$$\text{Allowed Time} = (\text{Normal Time}) \times \left(1 + \frac{\text{Allowances in }\%}{100}\right)$$

$$\text{Allowed Time} = 153.22 \times \left(1 + \frac{20\%}{100}\right)$$

Allowed Time = 183.864 sec

Since, the given policy allowances are 20 percent of the normal time.

$$\text{Standard Time} = (\text{AllowedTime}) \times \left(1 + \frac{\text{Policy Allowances in }\%}{100}\right)$$

$$\text{Standard Time} = (183.864) \times \left(1 + \frac{20\%}{100}\right)$$

$$= \mathbf{220.64 \text{ sec}}$$

Numerical 2: The following data refers to a sampling study of production of one component.

1. Duration of data collection = 5 days @ 8 hours per day
2. Number of operators = 10
3. Allowances given for the process = 15%
4. Production quantity in 5 days = 6000 components
5. Sampling data collected:

Days	1	2	3	4	5
No. of observations	230	240	200	180	225
Occurrence of activity	200	190	170	150	210

Calculate standard time of production of the component if average performance rating of the operator is 120% and the entire operation is manual.

Sol: No. of observations (N) = 1075

No. of observations (Np) = 920 .

Overall time per piece, $T_0 = \dfrac{\text{Total time worked}}{\text{No. of units produced}}$

$$= \dfrac{5 \times 8 \times 10 \times 60}{6000} \text{ min.}$$

$$= 40 \text{ min.}$$

Effective time per piece,

$$(T_e) = \dfrac{T_0 \times N_0}{N}$$

$$= \dfrac{40 \times 920}{175}$$

$$= 34.23$$

Normal time = Observed time x Rating

$$= 34.23 \text{ x } 1.2$$

$$= 41.07 \text{ min.}$$

Standard time = Normal time (1 + allowances)

$$= 41.07 (1 + 0.15)$$

$$= \textbf{47.24 min.}$$

Problems for Practice

Prob. 1. A job order shop has 12 general purpose machines. A work-sampling study has been designed to know the ineffective time of entire shop. The study conducted revealed that the unproductive time goes to the extent of 30%.

Compute the number of observations that are required to have the accuracy of 5% with the confidence level of 95%.

Prob. 2. A work-sampling study was conducted to establish the standard time for an operation. The observations of the study conducted are given below:

Total number of observations	= 160
Hand controlled work	= 14
Machine controlled work	= 106
Machine idle time	= 40
Average performance rating	= 80%
No. of parts produced	= 36
Allowance for personal needs and fatigue	= 10%
Study conducted for 03 days	
Available working hours/day	= 08 hrs.

Calculate the standard time per piece.

Prob. 3. The following data is available from time study on a job:

Representative time = 0.5 minutes

Rating = 125%

Allowances 15% of S.T

Determine Basic time and Standard time.

Prob. 4. The following data is available from time study on a job:

Representative time = 0.75 minutes, Rating = 110%, Relaxation Allowance =10%, Personal allowance=3%, Delay allowance=2%. All allowances are expressed in percentage of Normal time (N.T).

Determine the Standard Time.

Prob. 5. In a time study of a job by a worker whose rating is 90%, data obtained is as follows:

Observed time = 15 minutes

Personal needs allowance = 4% of basic time

Fatigue allowance = 2.5% of basic time

Contingency work allowance = 2% of basic time

Contingency delay allowance = 1% of basic time

Determine: (i) Basic time (ii) Work content and (iii)Standard time

4.8 Predetermined Motion Time Systems

A predetermined motion time system (PMTS) may be defined as a procedure that analyses any manual activity in terms of basic or fundamental motions required to perform it. Each of these motions is assigned a previously established standard time value and then the timings for the individual motions are synthesised to obtain the total time needed for performing the activity.

The main use of PMTS lies in estimating the time for the performance of a task before it is performed. The procedure is particularly useful to those organizsations which do not want troublesome performance rating to be used with each study.

Applications of PMTS are for

- Determination of job time standards
- Comparing the times for alternative proposed methods so as to find the economics of the proposals prior to the production run
- Estimation of manpower equipment and space requirements prior to setting up the facilities and starting of the production
- Developing tentative work layouts for assembly lines prior to their working in order to minimise the amount of subsequent rearrangement and rebalancing
- Checking direct time study results.

A number of PMTS are in use, some of which have been developed by individual organisations for their own use, while other organisations have developed and publicised for universal applications.

Some commonly used PMT systems are

(*a*) Work factor System (WFS)

(*b*) Motion Time Analysis (MTA)

(*c*) Basic Motion Time (BMT)

(*d*) Method Time Measurement (MTM)

(*a*) **Work Factor System (WFS):** Work Factor System got developed in Philadelphia in the year 1934. The two main names associated with Work Factor System were those of J.H. Quick and W.J. Shea.

Like MTM, Work Factor System, also relies on manuals containing time values for different elements (i.e. leg, trunk, foot, etc.) predetermined from high speed films of a large number of industrial operations.

Unlike MTM , Work Factor System

(*i*) is more accurate,

(*ii*) has a simple and easy procedure,

(*iii*) in addition to other aspects, it takes into account Mental Process Times,

(*iv*) considers some non-productive time also,

(*v*) has its standards for an experienced skilled worker, whereas MTM standards are based upon the performance of an average operator. Because of this reason, for the same job Work Factor Systems gives a smaller time as compared to MTM.

(*vi*) Has 1 Time Unit = 0.0001 minute.

(*b*) **Motion Time Analysis (MTA):** It was developed by A.B. Segur in 1924. It is considered as the fore runner of all other PMTS. It is based upon the fundamental motions, developed by Gilbreth, and it is called as 'therbligs'. Segur used the motion pictures, micro- motion analysis, and kymograph to establish the time values for various therbligs. It is used to estimate time values of human motions, by dividing them into 'therbligs' and building them up by adding the time values already developed.

(c) **Basic Motion Times (BMT):** This was developed by Palph Presgrave, Bailey and their associates of J.B. Woods and Gordon Limited of Toronto during 1945-1951. This system was developed through laboratory study and experimentation. It has the following concepts:

  (a) **Basic motion:** The basic motion is defined as a motion in which the body at rest moves and again comes to rest; the same is depicted in figure 4.10.

  (b) **Categories of basic motion:** Basic motions can be classified into three categories; A, B, and C depending upon the nature of stopping of the basic motion, and further into two sub-categories based upon the visual direction required.

   (i) Basic motion 'A' : Here the motion is stopped without muscular efforts or control

   (ii) Basic motion 'B': Here the motion is stopped with muscular control

   (iii) Basic motion 'C': Here the motion is slowed down by muscular control, i.e., before the object or tool is grasped or positioned.

(d) **Method Time Measurement (MTM):** It can be defined as the procedure that analyses any manual operation or method into basic motions required to perform it. It assigns a pre determined time standard to each motion. These time standards are determined by the nature of the motion and the conditions under which they are made. MTM was developed by Maynard, Stegmerten and Schwab during the Second World War. This is known to be the most popular method over the others. Since significant work has been made in each and every field of human activity: therefore, special types of MTMs have been developed for predetermining the time in maintenance work as well as for the measurement of non-repetitive indirect work. The MTM international Directorate has been setup and under its guidance various MTM associations have been at the national level, with a view to protect MTM against incorrect useage, coordinate research in the field of MTM and supervise the correct training of instructors and practitioners through approved courses.

Important considerations which may be made while selecting a PMT system for application to a particular industry are

- **Cost of Installation:** This consists mainly of the cost of getting an expert for applying the system under consideration.
- **Application Cost:** This is determined by the length of time needed to set a time standard by the system under consideration.
- **Performance Level of the System:** The level of performance embodied in the system under consideration may be different from the normal performance established in the industry where the system is to be used. However, this problem can be overcome by 'calibration' which is nothing but multiplying the times given in the PMT tables by the some constant or by the application of an adjustment allowance.
- **Consistency of Standards:** Consistency of standards set by a system on various jobs is a vital factor to consider. For this, the system can be applied on a trial basis on a set of operations in the plant and examined for its consistency in thus obtained operation times.
- **Nature of Operation:** Best results are likely to be achieved if the type and nature of operations in the plant are similar to the nature and type of operations studied during the development of the system under consideration.

Advantages and Limitations of using PMT systems

Advantages

Compared to other work measurement techniques, all PMT systems claim the following advantages:

- There is no need to actually observe the operation when it is running. This means the estimation of time to perform a job can be made from the drawings even before the job is actually done. This feature is very useful in production planning, forecasting, equipment selection, etc.

- The use of PMT eliminates the need of troublesome and controversial performance rating. For the sole reason of avoiding performance rating, some companies use this technique.

- The use of PMT forces the analyst to study the method in detail. This sometimes helps to further improve the method.

- A bye-product of the use of PM Times is a detailed record of the method of operation. This is advantageous for the installation of method, for instructional purposes, and for detection and verification of any change that might occur in the method in future.

- The PM Times can be usefully employed to establish elemental standard data for setting time standards on jobs done on various types of machines and equipments.

- The basic times determined with the use of PMT system are relatively more consistent.

Limitations

There are two main limitations in the use of PMT system for establishing time standards. These are: (i) it can be applied to only manual contents of a job, and (ii) it needs trained personnel. Although PMT system eliminates the use of rating, quite a bit of judgment is still necessarily exercised at different stages.

Summary

This chapter deals with the concepts and objectives of Work Measurement. It also describes the various techniques of Work Measurement. Besides, it highlights the methods of rating, allowances and determination of Normal Time and Standard Time. It also covers the applications of various types of Predetermined Motion Time Study, like Work Factor System, Motion Time Analysis, Basic Motion Time and Method Time Measurement.

$\boxed{5}$

Ergonomics

5.1 Introduction

Ergonomics is a science that includes different branches, such as engineering, anatomy, the man and machine systems, physiology and psychology. A person having experience in any of these areas will be able to work effectively on the ergonomic aspects related with the different design equipments. Such designed systems, which can be used by the human beings effectively with putting minimum effort and cause the maximum comfort in their use and application fall in category of the Ergonomics

Ergonomics is a science which deals with the relationship of man with machine and his working environment. It takes care of factors governing physical and mental strain that a worker goes through. Ergonomics consists of two Greek words 'Ergo' (meaning work), and 'Nomos' (meaning 'Natural Laws'). It can also be termed as 'Human Engineering'. Ergonomics (or Human Engineering) is defined by I.L.O. (International Labour Organisation) as "the application of human biological sciences in conjunction with engineering sciences to the worker and his working environment so as to obtain maximum satisfaction for the worker which, at the same time, enhances productivity".

In other words, Ergonomics may be defined as a multidisciplinary science comprising subjects like anatomy, psychology, physiology, sociology, engineering, anthropology, physics, medicine and statistics to ensure that designs of working implements complement the strength and abilities of the people working on them and minimise the effects of their limitations. The task of Ergonomics is to develop such conditions for workers, which are necessary to reduce their physical workload, to improve their working postures, facilitate instrument handling, and thus improve the quality of labour put in by reducing fatigue, maximising the efficiency of production operators and minimising human errors. Rather than expecting people to adapt to a design that forces them to work in an uncomfortable, stressful or dangerous manner, Ergonomists and human factors specialists seek to understand how a product, workplace or system can be designed to suit the people who need to use it.

Ergonomics provides the guiding principles and specifications according to which tools, machines, work procedures and workplaces are designed for safe and efficient use. The efficiency of a machine depends on the ability of the worker to control it effectively and accurately. The fact that workers are able to operate in poorly designed workplaces does not mean that it is the most efficient method of production; workers should be able to operate machines in the least stressful way. Ergonomics aims to create safe, comfortable and productive work spaces suitable to human abilities and limitations taking the individual's body size, strength, skill, speed, sensory abilities (vision, hearing), and even attitudes into consideration.

5.2 History of Development

Ergonomics emerged as a scientific discipline in the 1940s as a consequence of the growing realisation that as technical equipments became increasingly complex, not all of the expected benefits could be delivered if people were unable to understand and use the equipment to its full potential.

Initially, these issues were most evident in the military sector where high demands were placed on the physical and cognitive demands of the human operator. As the technological achievements of World War II were transferred to civilian applications, similar problems of disharmony between the people and equipments were encountered, resulting in poor user-performance and an increased risk of human error. The analysis of poor performance, of what came to be known as man-machine systems (now human-machine systems), provided a growing body of evidence which could be linked to difficulties faced by human operators. This motivated the senior academics, and military physiologists and psychologists to further investigate the nature of interaction of people with their equipments and environment. Although the early focus was on work environments, the importance of ergonomics has been increasingly recognised in many spheres, including the design of consumer products, such as cars and computers.

In 1949, at a meeting of distinguished physiologists and psychologists at The Admiralty, the term Ergonomics was coined from the Greek roots (Ergo and Nomos). Later that year this same body of scientists, together with some like-minded colleagues, formed the Ergonomics Research Society (ERS), which became the first such professional body in the world.

In the following sixty-two years, the ERS has evolved to represent the current discipline, both in the United Kingdom and internationally. In 1977 the ERS became the Ergonomics Society (ES) in recognition of the increasing focus on the professional application and practice of Ergonomics that stemmed from the ever-increasing theoretical and research base. The ES became a Registered Charity (number 292401) and a Company limited by guarantee (Company number 1923559) in 1985.

As the discipline evolved, some variations in terminology arose in different countries. In the USA the term human factors took on the same meaning as ergonomics in the UK. Although the two terms have remained synonymous to professionals, popular usage has recently accorded different shades of meaning to them. Consequently, human factors may be considered to imply the cognitive areas of the discipline (perception, memory, etc.,) whereas ergonomics may be used more specifically to refer to the physical aspects (workplace layout, light, heat, noise, etc.). In 2009, following a vote by the membership and approval from Companies House, the ES was renamed the Institute of Ergonomics and Human Factors (IEHF) to reflect the popular usage of both terms and to emphasise the breadth of the discipline.

In 2014, the discipline's importance was recognised officially by the award of a Royal Charter to the Institute. This allows us to confer Chartered status on those members who fulfill certain criteria. This includes having a high level of qualification and experience and being able to demonstrate continuing professional development. At the end of 2014, the Institute had 294 members who were eligible for Chartered status, with many more about to become so. They are the first such 'Chartered Ergonomists and Human Factors Specialists' in the world. The Charter and its accompanying governing documents were unanimously accepted by the membership at an Extraordinary General Meeting in November. The Institute changed its name once again to the Chartered Institute of Ergonomics & Human Factors.

5.3 Objectives

1. To optimise the integration of man and machine in order to increase productivity with accuracy. It involves the design of:

    (*a*) A workplace suitable for the worker

    (*b*) Machinery and controls, so as to minimise mental and physical strain on the worker to enable improvement in efficiency

    (*c*) A favourable environment for performing the task most effectively

    (*d*) Task and work organisation

2. To take care of the factors governing physical and mental strain (*i.e.,* fatigue), so as to get maximum satisfaction for the worker which at the same time would enhance productivity?

3. Attempt to minimise the risk of injury, illness, accidents and errors without compromising productivity

4. To improve the design of machine at the initial design stage or later whenever the existing product or process is modified

    (*a*) Developing most comfortable conditions related to climate, lighting, ventilation and noise level

    (*b*) Reducing the physical workload

(*c*) Improving working postures and reducing efforts of certain movements

(*d*) Making the handling of machine levers and controls easy

(*e*) Increasing safety

5.4 Man-Machine Systems: Design, Characteristics and Classification

Human factors are a system concerned with the relationship between human beings and their workplace or work environment and machines. All man-machine systems are produced with some objective in view. This objective is always well defined and the system is designed so as to achieve the objective as successfully as possible. In view of this the operational functions of both the components and constituents, i.e., man and machine should be clearly defined. There is one another aspect of man-machine system which, though not strictly a part of it, affects the system performance to a great extent. This is the system environment or what we call working conditions. The proper integration of man and machine, which is beneficial for human operator and enhances the overall system performance, is the primary aim of the Ergonomics discipline.

Characteristics of Man-Machine System

- The man-machine system consists of man, machine and system environment.
- It is essentially artificial in nature and is specifically developed to fulfill some purpose or specific aim.
- It has specific inputs and outputs which are appropriately balanced.
- It is variable in size and complexity and is dynamic in performance.
- A sub-system of man-machine system interacts with and affects the other parts.
- The man-machine system becomes more efficient when inputs and out puts are adequately balanced.
- Environmental factors or system environment affects system performance.

Classification of Man-Machine Systems

Depending upon the size and complexity, man-machine systems are of following three types:

1. Manual Systems

They are essentially man-directed systems. These are flexible in nature and small in size. Simple tools and equipments are used, and the efficiency depends upon the human factor. A large variability is possible in a manual system as every worker may select different method to do the same job.

2. Mechanical Systems

They are more complex and inflexible in nature than the manual ones. The machine component is power-driven, and human activity includes information-processing, decision-making and controlling occasionally the semi-automatic systems; they have components which are well-integrated. This is the feature which renders these systems rather inflexible. An automobile or a machine tool operated by a driver or an operator are good examples of this category.

3. Automatic Systems

A complex system in which all operational functions are performed by automatic devices is known as automatic system. Operational functions include sensing information, processing it, making decisions and taking action. It is totally inflexible in nature and cannot be adapted to uses other than that the one for which it has been designed. The human element/component performs the jobs of monitoring, programming the function, maintenaning and up keep of the system. An automatic telephone exchange, a digital computer and an automatic screw of cutting machine are good examples of automatic systems. A perfectly reliable automatic system does not exist at present.

5.5 Introduction to Structure of Human Body

The sense organs of man make physical contact with their environment. Through their senses information is conveyed to the brain, such as the eyes, ears and nose. The stimulus has to be strong enough for the senses to detect and become aware of its presence in the environment. The 'absolute threshold' marks the difference between being aware and not being aware of a stimulus; this may vary at different times and under different conditions. A second threshold, termed as the 'difference threshold' refers to detectable differences between two stimuli that can be sensed by an individual. People's senses automatically adapt to various stimuli in different situations. But if the stimulus is constant and familiar the sense organs can become insensitive to it.

5.5.1 Features of Human Body

Human body is composed of elements like hydrogen, oxygen, carbon, calcium and phosphorus. These elements reside in trillions of cells and non-cellular components of the body.

The content, acidity and composition of the water inside and outside of cells is carefully maintained. The main electrolytes in body water outside the cells are sodium and chloride, whereas potassium and other phosphates are found within the cells.

Cells

The body contains trillions of cells— the fundamental unit of life. At maturity, there are roughly 30-37 trillion cells in the body, an estimate arrived at by totalling the cell numbers of all the organs of the body and cell types. The body is also host to about the same number of non-human cells as well as multicellular organisms which reside in the gastrointestinal tract and on the skin.

Tissues

The body consists of many different types of tissues, defined as cells, that have a specialised function. The study of tissues is called histology; the tissues can be seen with a microscope. The body consists of four main types of tissues - lining cells (epithelia), connective tissues, nervous tissues and muscle tissues. Cells that lie on surfaces exposed to the outside world, or within the gastrointestinal tract (epithelia) or internal cavities (endothelium) are in numerous shapes and forms — from single layers of flat cells, to cells with small beating hair-like cilia in the lungs, to column-like cells that line the stomach. Endothelial cells are cells that line internal cavities including blood vessels and glands. Lining cells regulate what can and can't pass through them, protect internal structures, and function as sensory surfaces.

Organs

Organs, structured collection of cells with a specific function, sit within the body. Their examples include heart, lungs and liver. Many organs reside within the cavities inside the body. These cavities include the abdomen and pleura.

Circulatory system

The circulatory system comprises the heart and blood vessels (arteries, veins and capillaries). The heart propels the circulation of the blood, which serves as a "transportation system" to transfer oxygen, fuel, nutrients, waste products, immune cells and signaling molecules (*i.e.,* hormones) from one part of the body to another. The blood consists of fluid that carries cells in the circulation, including some that move from tissue to blood vessels and back, as well as the spleen and bone marrow.

Digestive system

The digestive system consists of the mouth including the tongue and teeth, esophagus, stomach, (gastrointestinal tract, small and large intestines, and rectum), as well as the liver, pancreas, gallbladder, and salivary glands. It converts food into small, nutritional, non-toxic molecules for distribution and absorption into the body.

Immune system

The immune system consists of the white blood cells, the thymus, lymph nodes and lymph channels, which are also part of the lymphatic system. The immune system provides a mechanism for the body to distinguish its own cells and tissues from outside cells and substances, and to neutralise or destroy the latter by using specialised proteins, such as antibodies, cytokines, and toll-like receptors among many others.

Eyes

Eyes operate like a camera catching (through the pupils) and refracting light (lens) and converting it into a picture (retina to the optic nerve). Eyes are susceptible to hazards, such as flying particles and irritating dusts, chemical or radiation damage and, in cases of inadequate lighting, eyestrain. Protection for the eyes can be achieved either with physical barriers that protect them against foreign objects, *e.g.,* safety glasses or by improving workplace and task-design so that they do not have to strain them selves too hard. This can be done by reducing glare and reflections, optimising workplace lighting and viewing object qualities, such as contrast, colour, size and shape.

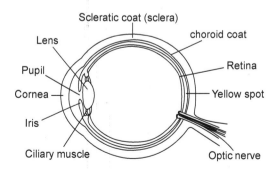

Fig. 5.1 Antomoy of a Human Eye

This includes the design of displays and printed materials. A small percentage of the population is colour blind. These people are usually men with varying degrees of red/green blindness. This can be critical for certain occupations, and while viewing visual displays involving these colours.

Ears

The ear and auditory system are more complicated than most people realise. It consists of the external ear, middle ear (separated from the external ear by the ear drum), inner ear and the central auditory pathways. Sound travels to the ear in waves. These are transmitted via the auricle (visible part) and external auditory canal through the eardrum to small bones in the middle ear that vibrates. From there the vibrations are transmitted to the inner ear and to the sensory cells of the cochlear that respond to particular frequencies or pitches. The cells transform damaged sound waves to nerve impulses that are transmitted to the brain. The cells and hairs can get damaged at times when exposed to loud noise.

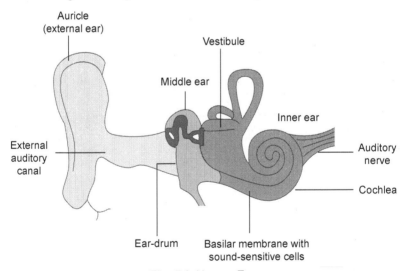

Fig. 5.2 Human Ear

Different kinds of noise affects hearing in different ways: the higher the pitch the worse the effect is; the clearer the tone the greater the hazard; the higher the intensity the greater the damage; the greater the length of exposure to damaging noise on a daily basis the greater is the risk of hearing loss. (See also Noise). The ears also contain the semicircular canals that are necessary for balance and body orientation.

Skin

The skin is the largest (1.4-2 square meters) visible part of an individual, and so is the body's largest organ. It protects tissues underneath from physical and chemical damage as well as protect the body from drying out, and abrupt changes in temperature. The skin contains: • sweat glands that help maintain an even body temperature, and fine blood vessels that assist in temperature control, nutrition and waste removal; nerve endings that act as sensory receptors of heat, cold, pain, pressure and touch; sebaceous glands that secrete substances to keep the skin supple and protect it from harmful bacteria. Exposure of the skin to some substances and physical agents, such as the sun, may cause skin irritation, non-allergic contact eczema and burning. Protection of the skin is achieved best through elimination of or isolation from the substances and agents, and less effectively with PPE.

Nose

The nose transmits sensations of smell and filters as well as alters the temperature of the air that an individual inhales. An individual's sense of smell adapts quickly to certain smells. However, some of these may tell a worker that there is a problem. Workers may need respiratory protection in environments where unpleasant or obnoxious smells cannot be eliminated. Dangerous, unnecessary and/or unpleasant smells care required to be controlled where the sense of smell works as an early detection monitor.

Taste

Taste buds are on the tongue and respond to the sensations of sweetness, saltish, bitterness and sour tastes.

5.5.2 Stress and Strain

The concept of fatigue and recovery at human work is closely related to the ergonomic concepts of stress and strain.

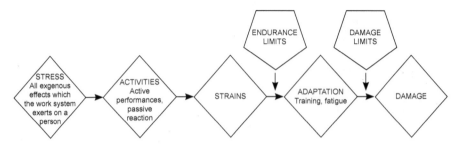

Fig. 5.3 Stress, Strain and Fatigue

Stress implies the sum of all parameters of work in the working system influencing people at work, which are perceived or sensed mainly over the receptor system, or which put demands on the effector system. The parameters of stress result from the work task (muscular work, non-muscular work and task-oriented dimensions and factors) and from the physical, chemical and social conditions under which the work has to be done (noise, climate, illumination, vibration, shift work, etc. situation-oriented dimensions and factors).

The intensity/difficulty, the duration and the composition (i.e., the simultaneous and successive distribution of these specific demands) of the stress factors results in combined stress, which all the exogenous effects of a working system exert on the person working. This combined stress can be actively coped with or passively put up with, specifically depending on the behaviour of the engaged person. The active case will involve activities directed towards the efficiency of the working system, while the passive case will induce reactions (voluntary or involuntary), which are mainly concerned with minimising stress. The relation between the stress and activity is decisively influenced by the individual characteristics and needs of the working person. The main factors of influence are those that determine performance and are related to motivation, and concentration and to disposition, which can be referred to as abilities and skills.

The stresses relevant to behaviour, which are manifest in certain activities, cause individually different strains. The strains can be indicated by the reaction of physiological or biochemical indicators (e.g., raising the heart rate) or it can be perceived. Thus, the strains are susceptible to "psycho-physical scaling", which estimates the strain as experienced by the working person. In a behavioural approach, the existence of strain can also be derived from an activity analysis. The intensity with which indicators of strain (physiological-biochemical, behavioristic or psycho-physical) react depends on the intensity, duration and combination of stress factors as well as on the individual characteristics, abilities, skills and needs of the working person.

Despite constant stresses the indicators derived from the fields of activity, performance and strain may vary over time (temporal effect). Such temporal variations are to be interpreted as processes of adaptation by the organic systems. The positive effects cause a reduction of strain/ improvement of activity or performance (e.g., through training). In the negative case, however, they will result in increased strain/reduced activity or performance (e.g., fatigue, monotony).

The positive effects may come into action if the available abilities and skills are improved in the working process itself, e.g., when the threshold of training stimulation is slightly exceeded. The negative effects are likely to appear if so-called endurance limits (Rohmert 1984) are exceeded in the course of the working process. This fatigue leads to a reduction of physiological and psychological functions, which can be compensated by recovery. When the process of adaptation is carried beyond defined thresholds, the employed organic system may be damaged so as to cause a partial or total deficiency of its functions. An irreversible reduction of functions may appear when stress is far too high (acute damage) or when recovery is impossible for a longer time (chronic damage). A typical example of such damage is noise-induced hearing loss.

5.5.3 Metabolism

It is a collection of chemical reactions that takes place in the body's cells. Metabolism converts the fuel in the food we eat into the energy needed to power everything we do, from moving to thinking to growing. Specific proteins in the body control the chemical reactions of metabolism, and each chemical reaction is coordinated with other body functions. In fact, thousands of metabolic reactions happen at the same time — all regulated by the body — to keep our cells healthy and working. Metabolism is a constant process that begins when we're conceived and ends when we die. It is a vital process for all life forms — not just humans. If metabolism stops, living things die.

Here's an example of how the process of metabolism works in humans, and it begins with plants:

- First, a green plant takes in energy from sunlight. The plant uses this energy and a molecule called chlorophyll (which gives plants their green color) to build sugars from water and carbon dioxide. This process is called **photosynthesis**, and you have probably learned about it in biology class.

- When people and animals eat the plants (or, if they're carnivores, they eat animals that have eaten the plants), they take in this energy (in the form of sugar), along with other vital cell-building chemicals. Then, the body breaks the sugar down so that the energy released can be distributed to, and used as fuel by the body's cells.

- After the food is consumed, molecules in the digestive system called **enzymes** break proteins down into amino acids, fats into fatty acids, and carbohydrates into simple sugars (e.g., glucose). Like sugar, amino acids and fatty acids can be used as energy sources by the body when needed.

- These compounds are absorbed into the blood, which carries them to the cells. In the cells, other enzymes act to speed up or regulate the chemical reactions involved with "metabolizing" the compounds. The energy from these compounds can be released for use by the body or stored in body tissues, especially the liver, muscles, and body fat.

In static work, muscle contraction does not produce visible movement; for example, in a limb. Static work increases the pressure inside the muscle, which together with the mechanical compression occludes blood circulation partially or totally. The delivery of nutrients and oxygen to the muscle and the removal of metabolic end-products from the muscle are hampered. Thus, in static work, muscles become fatigued more easily than in dynamic work.

The most prominent circulatory feature of static work is a rise in blood pressure. Heart rate and cardiac output do not change much. Above a certain intensity of effort, blood pressure increases in direct relation to the intensity and the duration of the effort. Furthermore, at the same relative intensity of effort, static work with large muscle groups produces a greater blood pressure response than does the work done using smaller muscles.

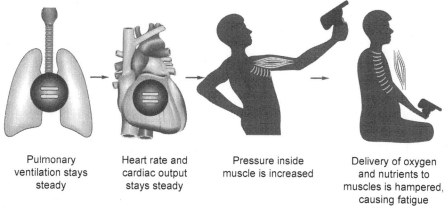

| Pulmonary ventilation stays steady | Heart rate and cardiac output stays steady | Pressure inside muscle is increased | Delivery of oxygen and nutrients to muscles is hampered, causing fatigue |

Fig. 5.4 Increase in Blood Pressure Due to Static Work

In principle, the regulation of ventilation and circulation in static work is similar to that in dynamic work, but the metabolic signals from the muscles are stronger and induce a different response pattern.

5.6 Measurement of Physiological Functions

Physiological measurement techniques are used to measure bodily variations either directly or indirectly; examples are measurements of heart rate, mean arterial pressure, and total lung capacity. Psychophysiology measures exist in three domains: reports, readings and behaviour.

5.6.1 Workload and Energy Consumption

Physical work can be done if a person has energy, as it requires the support of muscle contraction. The human body needs energy to sustain / maintain the basic functions of life even when there are no activities carried out at all. Lowest energy expenditure required to maintain the basic functions of life are called basal metabolism (basal metabolism) (Wickens et. al.).

Each individual has a different basal metabolic rates. Some of the factors that influence these differences are gender, age, and body weight. Based on the description, physical workloads can be classified into several levels based on energy expenditure. Some work physiology studies explain that the energy spent on a work done is directly proportional to the amount of oxygen consumption and heart rate (Wickens et al.). Dr. Lucien Brouha has made the workload classification table in physiological reactions to determine the severity of a job, as is shown in table 1 below:

Table 5.1 Level of Wok versus Energy Consumption, Hear Beat and Oxygen Consumption

Level of work	Energy expenditure		Heartbeat	Oxygen Consumption
	Kkal / menit	Kkal / 8jam	Beat / min	Liter / min
Unduly Heavy	>12.5	>6000	>175	<2.5
Very Heavy	10.0 – 12.5	4800 – 6000	150 – 175	2.0 – 2.5
Heavy	7.5 – 10.0	3600 – 4800	125 – 150	1.5 – 2.0
Moderate	5.0 – 7.5	2400 – 3600	100 – 125	1.0 – 1.5
Light	2.5 – 5.0	1200 – 2400	60 – 100	0.5 – 1.0
Very Light	<2.5	<1200	<60	<0.5

5.6.2 Bio-Mechanics

Bio-Mechanics is a discipline that approaches the study of the body as though it were solely a mechanical system: all parts of the body are likened to mechanical structures and are studied as such. The following analogies may, for example, be drawn:

- Bones: levers, structural members
- Flesh: volumes and masses
- Joints: bearing surfaces and articulations
- Joint linings: lubricants
- Muscles: motors, springs
- Nerves: feedback control mechanisms
- Organs: power supplies
- Tendons: ropes

- Tissue: springs
- Body cavities: balloons.

The main aim of biomechanics is to study the way the body produces force and generates movement. The discipline relies primarily on anatomy, mathematics and physics; related disciplines are anthropometry (the study of human body measurements), work physiology and kinesiology (the study of the principles of mechanics and anatomy in relation to human movement). While considering the occupational health of the worker, biomechanics helps to understand why some tasks cause injury and ill health. Some relevant types of adverse health effects are muscle strain, joint problems, back problems and fatigue.

Back strains and sprains and more serious problems involving the intervertebral discs are common examples of workplace injuries that can be avoided. These often occur because of a sudden particular overload, but may also reflect the exertion of excessive forces by the body over many years: such problems may occur suddenly or may take time to develop. An example of a problem that develops over time is "seamstress's finger". A recent description describes the hands of a woman who, after 28 years of work in a clothing factory, as well as sewing in her spare time, developed hardened thickened skin and an inability to flex her fingers (Poole 1993). (Specifically, she suffered from a flexion deformity of the right index finger, prominent Heberden's nodes on the index finger and thumb of the right hand, and a prominent callosity on the right middle finger due to constant friction from the scissors.) X-ray films of her hands showed severe degenerative changes in the outermost joints of her right index and middle fingers, with loss of joint space, articular sclerosis (hardening of tissue), osteophytes (bony growths at the joint) and bone cysts.

Inspection at the workplace showed that these problems were due to repeated hyperextension (bending up) of the outermost finger joint. Mechanical overload and restriction in blood flow (visible as a whitening of the finger) would be maximal across these joints. These problems develop in response to repeated muscle exertion in a site other than the muscle. Biomechanics helps to suggest ways of designing tasks to avoid these types of injuries or of improving poorly designed tasks. Remedies for these particular problems are to redesign the scissors and to alter the sewing tasks to remove the need for the actions performed.

5.6.3 Types of Movements of Body Members

Exion: movement in the sagittal plane that decreases the angle of the joint and brings two bones closer together

Extension: opposite of flexion; movement in the sagittal plane that increases the angle of the joint or distance between two bones or parts of the body

Hyperextension: extension greater than 180 degrees

Rotation: movement of a bone around its longitudinal axis

Abduction: moving a limb away in the frontal plane from the median plane of the body, spreading the fingers apart

Adduction: opposite of abduction; movement of a limb toward the body midline

Circumduction: a combination of all the movements, commonly seen in ball and socket joints where the proximal end of the limb is stationary while the distal end moves in a circle

Dorsiflexion: lifting the foot so the superior surface approaches the shin, standing on the heels

Plantar flexion: pointing the toes

Inversion: turning the sole of the foot medially

Eversion: turning the sole of the foot laterally

Supination: forearm rotation laterally so that the palm is facing anteriorly and the radius and ulna are parallel

Pronation: forearm rotation medially so that the palm faces posteriorly and the ulna and radius are crossed

Opposition: touching the thumb to other fingers.

5.6.4 Strength and Endurance

Muscular strength and endurance are two important parts of your body's ability to move, lift things and do day-to-day activities. Muscular strength is the amount of force you can put out or the amount of weight you can lift. Muscular endurance is how many times you can move that weight without getting exhausted (very tired).

Muscular strength and endurance are important for many reasons:

- Increase your ability to do activities like opening doors, lifting boxes or chopping wood without getting tired.
- Reduce the risk of injury.
- Help you keep a healthy body weight.
- Lead to healthier, stronger muscles and bones.
- Improve confidence and how you feel about yourself.
- Give you a sense of accomplishment.
- Allow you to add new and different activities to your exercise programme.
- Improving muscular strength and endurance
- There are many ways to improve muscular strength and endurance. A gym or fitness centre is a good place to go if you're interested in doing resistance training (also called strength training, weight training or weight lifting). This involves working a muscle or group of muscles against resistance to increase strength and power.

5.6.5 Speed of Movements

In immediate postwar industry the overriding objective, shared by ergonomics, was greater productivity. This was a feasible objective for ergonomics because so much industrial productivity was determined directly by the physical effort of the workers involved, like speed of assembly, rate of lifting and movement determined the extent of output. Gradually, mechanical power replaced human muscle power. More power, however, leads to more accidents on the simple principle that an accident is the consequence of power in the wrong place at the wrong time. When things are happening faster, the potential for accidents is further increased. Thus the concern of industry and the aim of ergonomics gradually shifted from productivity to safety. This occurred in the 1960s and early 1970s. About and after this time, much of manufacturing industry shifted from batch production to flow and process production. The role of the operator shifted correspondingly from direct participation to monitoring and inspection. This resulted in a lower frequency of accidents because the operator was more remote from the scene of action, but sometimes in a greater severity of accidents because of the speed and power inherent in the process.

5.7 Anthropometry

Anthropometry is a fundamental branch of physical anthropology. It represents the quantitative aspect. A wide system of theories and practice is devoted to define methods and variables to relate the aims in different fields of application. In the fields of occupational health, safety and ergonomics anthropometric systems are mainly concerned with the body build, composition and constitution, and with the dimensions of the human body's interrelation to workplace dimensions, machines, the industrial environment and clothing.

Anthropometric Variables

An anthropometric variable is a measurable characteristic of the body that can be defined, standardised and referred to a unit of measurement. Linear variables are generally defined by landmarks that can be precisely traced on the body. Landmarks are generally of two types: skeletal-anatomical, which maybe found and traced by feeling bony prominences through the skin, and virtual landmarks that are simply found as maximum or minimum distances using the branches of a caliper. Anthropometric variables have both genetic and environmental components and may be used to define individual and population variability. The choice of variables must be related to the specific research purpose and standardised with other research in the same field, as the number of variables described in the literature is extremely large, up to 2,200 have been described for the human body.

Anthropometric variables are mainly linear measures, such as heights, distances from landmarks with subject standing or seated in standardised posture; diameters, such as distances between bilateral landmarks; lengths, such as distances between two different landmarks; curved measures, namely arcs, such as distances on the body surface between two landmarks; and girths, such as closed all-around measures on body surfaces, generally positioned at least at one landmark or at a defined height. Other variables may require special methods and instruments. For instance, skinfold thickness is measured by means of special constant pressure calipers. Volumes are measured by calculation or by immersion in water. To obtain full information on body surface characteristics, a computer matrix of surface points may be plotted using biostereo metric techniques.

5.8 Design of Seat and Workplace

The best way to reduce pressure in the back is to be in a standing position. However, there are times when you need to sit. While sitting, the main part of the body weight is transferred to the seat. Some weight is also transferred to the floor, back rest, and armrests. Where the weight is transferred is the key to a good seat design. When the proper areas are not supported, sitting in a seat all day can put unwanted pressure on the back causing pain. The lumbar (bottom five vertebrate in the spine) needs to be supported to decrease disc pressure. Providing both a seat back that inclines backwards and has a lumbar support is critical to prevent excessive low back pressures. The combination which minimises pressure on the lower back is to have a backrest inclination of 120 degrees and a lumbar support of 5 cm. The 120 degrees inclination means the angle between the seat; and the backrest should be 120 degrees. The lumbar support of 5 cm means the chair backrest supports the lumbar by sticking out 5 cm in the lower back area. One drawback in creating an open body angle by moving the backrest backwards is that it takes one's body away from the tasking position, which typically involves leaning inward towards a desk or table. One solution to this problem can be found in the kneeling chair. A proper kneeling chair creates the open body angle by lowering the angle of the lower body, keeping the spine in alignment and the sitter properly positioned to task. The benefit of this position is that if one leans inward, the body angle remains 90 degrees or wider.

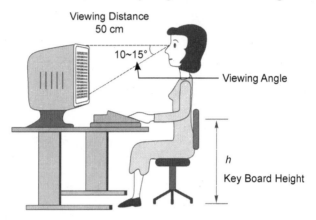

Fig. 5.5 Design of Seat and Workplace

One misconception regarding kneeling chairs is that the body's weight bears on the knees, and thus users with poor knees cannot use the chair. This misconception has led to a generation of kneeling chairs that attempt to correct this by providing a horizontal seating surface with an ancillary knee pad. This design wholly defeats the purpose of the chair. In a proper kneeling chair, some of the weight bears on the shins, not the knees, but the primary function of the shin rests (kneerests) is to keep one from falling forward out of the chair. Most of the weight remains on the buttocks. Another way to keep the body from falling forward is with a saddle seat. This type of seat is generally seen in some sit-stand stools, which seek to emulate the riding or saddle position of a horseback rider, the first "job" involving extended periods of sitting. Another way to reduce lumbar disc pressure is to use armrests. They help by putting the force of your body not entirely on the seat and back rest, but by putting some of this pressure on the armrests as well. Armrests need to be adjustable in height to ensure that the shoulders are not overstressed.

5.8.1 Steps in the Workplace Design

In the workstation design and implementation process, there is always an initial need to inform the users, and to organise the project so as to allow their full participation. This would increase the chance of full employee acceptance of the final results. A treatment of this goal is not within the scope of the present treatise which concentrates on the problem of arriving at an optimal solution for the physical design of the workstation, but the designing process nonetheless allows the integration of such a goal. In this process, the following steps should always be considered:

- Collection of user-specific demands
- Prioritising demands
- Transfer of demands into (a) technical specifications, and (b) specifications in user terms
- Iterative development of the workstation's physical layout
- Physical implementation
- Trial period of production
- Full production
- Evaluation and identification of resting problems.

The focus here is on steps one to five. Many times, only a subset of all these steps is actually included in while designing workstations. There are various reasons for this. If the workstation is of a standard design, such as in some Visual Display Unit (VDU) working situations, some steps may duly be excluded. However, in most cases the exclusion of some of the steps listed would lead to a lower quality of workstation than what can be considered acceptable. This may be the case when financial or time constraints are too severe, or when there is a sheer neglect of things due to the lack of knowledge or insight at the management level.

5.8.2 Other Workplace Risk Factors

The risk factors addressed by industrial ergonomics are a partial list of hazards present in a work-setting. Others include:

- Job stress
- Job invariability
- Cognitive demands
- Work organisation
- Workload
- Working hours (shift work, overtime)
- Displays and control panels
- Slip and falls
- Fire
- Electrical exposures
- Chemical exposures
- Biological exposures
- Ionising radiation
- Radiofrequency/microwave radiation
- Professionals such as industrial hygienists, human factors analysts, safety engineers, occupational medicine physicians, and occupational nurses evaluate and control these other risks. The ergonomist must recognise the skills and capabilities of these individuals. A close working relationship among these is essential for preparing an almost perfect workplace where prime considerations are health and sagely.

Certain features at of the work-setting have been associated with injury. These are called physical risk factors and include the following:

Posture: Posture is the position that a body takes while performing work activities. Awkward postures are associated with increased risks of injury. It is generally considered that the more a joint deviates from its neutral (natural) position, the greater is the risk of injury. Posture issues are caused by work methods (bending and twisting to pick up a box;

bending the wrist to assemble a part) or workplace dimensions (extended reach to obtain a part from a bin at a high location; kneeling in the storage bay of an airplane because of confined space while handling luggage).

Specific postures have been associated with injury. For example,

- Wrist
- Flexion/extension (bending up and down)
- Ulnar/radial deviation (side bending)
- Shoulder
- Abduction/flexion (upper arm positioned out to the side or above shoulder level)
- Hands at or above shoulder height
- Neck (cervical spine)
- Flexion/extension or bending the neck forward and to the back
- Side bending as when holding a telephone receiver on the shoulder
- Low back
- Bending at the waist, twisting

Force: Task forces can be viewed as the effect of an exertion on internal body tissues (e.g., compression on a spinal disc from lifting, tension within a muscle/tendon unit from a pinch grasp), or the physical features associated with an object(s) external to the body (e.g., weight of a box, pressure required to activate a tool, pressure necessary to snap two pieces together). Generally, the greater is the force, the higher is the degree of risk. High force has been associated with the risk of injury at the shoulder/neck (Berg et al.), the low back (Herrin et al.), and the arm/wrist/hand (Silver stein et al). It is important to note that the relationship between force and degree of injury risk is modified by other work risk factors, such as posture, acceleration/velocity, repetition, and duration.

Better analysis tools recognise the interrelationship of force with other risk factors relative to overall task risk. Five additional force-related injury risk conditions have been extensively studied by researchers and ergonomists. They are not "rudimentary" risk factors. Rather, they are a workplace conditions that present a combination of risk factors with force being a significant component. Their common appearance in the workplace and strong association with injury prompts their introduction here.

Static Exertion: Although defined in a variety of ways, static exertion generally means the performance of a task from one postural position for an extended duration. The condition is a combination of force, posture, and duration. The degree of risk is in proportion to the combination of the magnitude of the external resistance, awkwardness of the posture, and duration.

Grip: A grip is the conformity of the hand to an object accompanied by the application of exertion usually to manipulate the object. Hence, it is a combination of a force with a posture. Grips are applied to tools, parts, and other physical objects in the work-setting during task performance. To generate a specific force, a pinch grip requires a much greater muscle exertion than a power grip (object in the palm of the hand). Hence, a pinch grip has a greater likelihood of creating injury. The relationship between the size of the hand and the size of the object also influences risk of injury. Grant et al. found reduced physical exertion when the handle was one cm less than the subjects' grip-diameter.

Contact Trauma

Two types of contact trauma are:

(*a*) Local mechanical stress generated from sustained contact between the body and an external object, such as the forearm against the edge of a counter.

(*b*) Local mechanical stress generated from shock impact, such as using the hand to strike an object.

The degree of injury risk is in proportion to the magnitude of force, duration of contact, and sharpness of external object.

Gloves

Depending on material, gloves may affect the grip force generated by a worker for a given level of muscular exertion. To achieve a certain grip force while wearing gloves, a worker may need to generate greater muscular exertion than when not wearing gloves. Greater force is associated with increased risk of injury.

Bulky Clothes

Bulky clothes, used to protect the worker from cold or other physical elements, may increase the muscle effort required to perform tasks.

Velocity

Angular velocity/angular acceleration are the speed of body-part motion and the rate of change of speed of body-part motion, respectively. Marras and Schoenmarklin found a mean wrist flexion/extension acceleration of 490 deg/sec sec in low risk jobs and acceleration of 820 deg/sec in high risk jobs. Marras et al. associated trunk lateral velocity and trunk-twisting velocity with medium and high-risk occupationally-related low back disorder.

Repetition

Repetition is the time quantification of a similar exertion performed during a task. A warehouse worker may lift and place on the floor three boxes per minute; an assembly worker may produce 20 units per hour. Repetitive motion has been associated with injury and worker discomfort. Generally, the greater is the number of repetitions, the higher is the degree of risk. However, the relationship between repetition and degree of injury risk is modified by other risk factors, such as force, posture, duration, and recovery time. No specific repetition threshold value (cycles/unit of time, movements/unit of time) is associated with injury.

Duration

Duration is the time quantification of exposure to a risk factor. Duration can be viewed as the minutes or hours per day the worker is exposed to a risk. Duration also can be viewed as the years of exposure to a risk factor or a job characterised by a risk factor. In general, the greater is the duration of exposure to a risk factor, the greater is the degree of risk.

Recovery time

Recovery time is the time quantified for rest, performance of low stress activity, or performance of an activity that allows a strained body area to rest. Short work pauses have reduced perceived discomfort, and rest periods between exertions have reduced performance decrement. The recovery time needed to reduce the risk of injury increases as the duration of risk factor increases. Specific minimum recovery times for risk factors have not been established.

Heavy dynamic exertion

The cardiovascular system provides oxygen and metabolites to muscle tissues. Some tasks require long-term/repetitive muscle contractions such as walking great distances, heavy-carrying, and repeat-lifting. As physical activity increases, muscles demand more oxygen and metabolites. The body responds by increasing the breathing rate and heart rate.

When muscle demand for metabolites cannot be met (metabolic energy expenditure rate exceeds the body's energy producing and lactic acid removal rate) physical fatigue occurs. When this happens in a specific area of the body (e.g., shoulder muscle from repeat or long-term shoulder abduction), it is termed localized fatigue and is characterised by tired/sore muscles. When this happens to the body in general (from long-term heavy-carrying/lifting/climbing stairs), it is termed whole body fatigue and may produce a cardiovascular accident. Also, high heat from the environment can cause an increase in heart rate through body cooling mechanisms. Therefore, for a given task, metabolic stress can be influenced by environmental heat.

Segmental vibration (Hand-Arm vibration)

Vibration applied to the hand can cause a vascular insufficiency of the hands/fingers (Raynaud's disease or vibration white finger). Also, it can interfere with sensory receptor feedback leading to increased hand-grip force to hold the tool. Further, a strong association has been reported between carpal tunnel syndrome and segmental vibration.

5.9 Visual Display Design

There are three basic types of visual displays:

   (*a*) The check display indicates whether or not a given condition exists (for example a green light indicates normal function).

   (*b*) The qualitative display indicates the status of a changing variable or its approximate value, or its trend of change (for example, a pointer moves within a "normal" range).

   (*c*) The quantitative display shows exact information that must be ascertained (for example, to find a location on a map, to read text or to draw on a computer monitor), or it may indicate an exact numerical value that must be read by the operator (for example, a time or a temperature).

Design guidelines for visual displays are:

   (*a*) Arrange displays in such a way that the operator can locate and identify them easily without unnecessary searching. (This usually means that the displays should be in or near the medial plane of the operator, either below or at eye height.)

   (*b*) Group displays can be arranged functionally or sequentially so that the operator can use them easily.

   (*c*) Make sure that all displays are properly illuminated, coded and labelled according to their function.

   (*d*) Use lights, often coloured, to indicate the status of a system (such as ON or OFF) or to alert the operator that the system, or a subsystem is inoperative and that special action must be taken. Common meanings of light colours are: Flashing red indicates an emergency condition that requires immediate action. An emergency signal is most effective when it combines sounds with a flashing red light.

5.9.1 Factors Influencing Visual Display Effectiveness

Physical Location

The positioning of a visual display is of key importance in determining its effectiveness. The 'textbook' example of the 'user-interface' may conjure up an image of an individual setting in front of a computer screen. Under these circumstances, the optimised location of it should not present itself as a major problem. However, location of a display that is out of the main field of view or requires head or body movements, illustrates the problem of the wide variety of physical locations that a display can have. Information about the way visual acuity drops off as an object's retinal location moves from the fovea to the periphery can be used to site displays based upon the relative importance of the information conveyed, the size of the informational content and the time course of the signal. In the present context, the assumption can be made that the physical location of the display is optimised for the purpose for which it is required. If a user has to make frequent references to the current time, the positioning of the clock within his/her normal field of view would be self-evident. For a user who makes infrequent use of the current time a more peripheral location can be used.

Display Arrangements

The example of clock used above is not an appropriate one when it comes to thinking about the arrangement of multiple displays. In many advanced applications, it is likely that a number of displays will need to be grouped in close proximity to one another. This will be the case with motor car or aircraft instruments, etc. Similarly, items appearing on a computer screen for a set of tasks should all fall into the general field of view of the user. At the same time, they will need to be clearly distinguishable from one another. Examination of some display configurations may lead to the belief that such arrangements are governed largely by random assignments or aesthetics. It is, however, possible to collect hard data on the optimal arrangement of a set of displays (or indeed controls). Items of primary importance should be placed in the central field of view, while those of less importance must be on the periphery. Importance should be decided by an analysis of the tasks or by expert-rating.

Lighting Conditions

The extent to which the display is usable will depend upon the extent to which there is sufficient light falling on or emitted by the display.

In the case of warning lights and computer displays, they have light sources within themselves, and although adjustments may be made for 'brightness', there should be no problem over viewing them, even in a darkened room. Supplementary lighting may be required for other displays. There is usually background lighting for car instruments, which in turn can be adjusted for brightness, but for many displays, the natural or artificial light within the workspace is what will determine whether they can be viewed or not. As well as providing sufficient light for effective display, light sources can in turn give rise to problems. Variability in lighting from external sources may produce reflections when instruments are covered in glass. Contrast between the display and the external environment are also important. Dashboard instruments must remain legible by night and day. Internal lighting must be sufficient to negate the reduction of visual acuity in low light conditions at the same time as preserving adaptation or the 'night vision' required to drive in the dark. Bright lights may also cause problems.

Static versus Dynamic Displays

Another obvious manner in which displays can differ from one another is the extent to which they function purely as a static source, typically in the case of notices, signs, labels and instructions and the extent to which they represent dynamically changing features, typically called parameters. Analogue clocks display dynamic changes though it may be difficult to see the hands moving. Other displays associated with processes and changing states are much more dynamic in nature. Sometimes the changes are discrete as in the changing indication of how many miles/kilometers a car has travelled. Sometimes they are continuous in the way that the speedometer needle moves up and down in relation to changing speed. The complexity of a modern aircraft cockpit reflects the number of parameters that are being simultaneously measured and displayed for the pilot's attention. An important consideration with these complex displays is the limit on the number of items that humans can simultaneously process (cognitive psychologist have traditionally referred to this limit as the 'attentional-bottleneck').

5.10 Environmental Risk Factors

Heat Stress

Heat stress is the total heat load the body must accommodate. It is generated externally from environment temperature and internally from human metabolism. Excessive heat can cause heat stroke, a condition that can be life threatening or result in irreversible damage. Less serious conditions associated with excessive heat include heat exhaustion, heat cramps and heat-related disorders (e.g., dehydration, electrolyte imbalance, loss of physical/mental work capacity).

Cold Stress

Cold stress is the exposure of the body to cold such that there is a lowering of the deep core temperature. Systemic symptoms that a worker can present when exposed to cold include shivering, clouded consciousness, extreme pain, dilated pupils, and ventricular fibrillation. Cold can also reduce hand-grip strength and coordination. As mentioned earlier in the section on Force, bulky clothes and gloves used to protect the worker from cold exposure can increase the muscle effort required to perform tasks.

Whole Body Vibration

Exposure of the whole body to vibration (usually through the feet/buttocks when riding in a vehicle) has some support as a risk for injury. Boshuizen found the prevalence of reported back pain to be approximately 10 percent higher in tractor drivers than in workers not exposed to vibration, and the prevalence of back pain increased with vibration dose. Dupuis reported that operators of earth-moving machines with at least 10 years of exposure to the whole body vibration showed lumbar spine morphological changes earlier and more frequently than the non-exposed people.

Lighting

With industrialisation, the trend regarding lighting has been to provide a higher lighting level. This has proven hazardous within certain work-settings, such as in offices in which problems with glare and eye symptoms have been associated with levels above 1000 lux as suggested by Grandjean. Barreiros and Carnide found differences in visual functions over the course of a workday among **Video Display Terminal** (VDT) operators and money changers who worked in badly lighted environments. The current recommended trend in office lighting is to have low background lighting (from 300 to 700 lux) coupled with non-glare task lighting which can be controlled with a rheostat. This is consistent with Yearout and Konz's findings of operator preference regarding lighting. Work that requires high visual acuity and contrast sensitivity needs high levels of illumination. Fine and delicate work should be illuminated at 1,000 to 10,000 lux.

Noise

Noise means unwanted sound. In an industrial setting, it may be continuous or intermittent and present in various ways (bang of a rifle, clatter of a pneumatic wrench and whirl of an electric motor).Exposure to noise can lead to temporary or even permanent deafness, tinnitus, paracusis, or speech misperception. The louder the noise and greater its duration, the higher is the risk to hearing. Also, noise well below thresholds that cause hearing loss may interfere with the ability of some people to concentrate.

Air Quality

The next element of the work environment, which has impact on employee productivity, is air quality. Poor air quality can raise a negative impact on employees' health in the form of respiratory problems, headaches, and fatigue, which in the long period would reduce productivity. The air quality contains four factors and that are: temperature, humidity, ventilation, and cleanliness.

High Temperature Levels

Employee lethargy and tiredness as a result of increased body temperature decrease the possible efficiency.

Low Temperature Levels

Low Temperature Levels decrease the efficiency due to cooler body heat and shivering.

High Humidity

This in itself may not be a direct problem, but it does increase our susceptibility to high temperature levels as evaporation of body sweat is impeded.

Low Humidity

Levels have intolerable effect on our ability to breathe and swallow without discomfort as our mouths and noses can become dry due to an increased level of evaporation in the surrounding environment. A comfortable work environment of a building or room in which workers can do their work properly should be clean, have proper range of temperature, enough ventilation, and adequate humidity. Too little humidity level may cause magnetic tapes and disks to stick during processing operations, whereas too much humidity produces condensation on the electronic parts of the equipment and causes short circuiting.

Ventilation

Ventilation is important to control the dust, fumes, gases, aerosols, climate and thermal comfort factors. Exposure to different types of dust can result in fibrosis of the lung, allergic reactions and asthma attacks. Various vapours, gases and aerosols have the ability to cause respiratory and skin damage. Extreme heat can reduce the concentration of the worker and demotivate him, and may also cause a number of heat-related issues. Extreme heat can also reduce tolerance to chemical and noise exposure, and increase the risk of heart attacks.

After the temperature level in an office has been set-up properly suiting to the favourable level of humidity, the air in the office still needs to be circulated, otherwise it can increase the temperature, which in turn may cause discomfort. Air circulation is also essential as it can prevent workers' from inhaling inadequate air. Moreover, smoking must be prohibited in the office. Some small offices still use electric fans to make sure that the air is circulated well.

Air cleanliness is also becoming one of the prime considerations for the office management. Due to technological advances, many devices have been developed to clean the air. These devices free the air from germs, dust, and dirt.

Office Furniture

Office furniture comprises desks, chairs, the filing system, shelves, drawers, etc. All these components have a specific role to play in the proper functioning of any office and the productivity improvement of the employees. One of the most important aspect to be considered while buying office furniture is to ensure whether it is Ergonomic or not. Ergonomics of office furniture is important because an employee has to work with them for the entire time that he is in office; if they are uncomfortable and not user-friendly, employees' working style and performance get hampered considerably, which in turn affect the overall productivity of the organization concerned. Non-ergonomic office furniture can also lead to health problems of employees, which well again have an adverse effect on the productivity.

Ergonomic office furniture ensures that each employee gets comfortable with the things around him, like desks, chairs, computer alignment and even environmental factors. If all factors surrounding the employee are ergonomically suitable, he/she will be comfortable and remain motivated to give his best. These days' organisations consult, and even employ ergonomic experts to advice people on how to improve their office ergonomics to make it a better wore-palce. Having ergonomic office furniture reduces the chances of any risk injury. They are designed in such a manner that reduce the possibility of any accidents in the workplace. Office furniture like desks can be designed to give greater leg room and adequate support to the elbows while working on the computer. The positioning of the computer monitor and the mouse should also be adequate, so that the user does not have to strain his vision to view and stretch himself/herself uncomfortably to reach them. Proper office furniture helps the organisation tremendously in increasing its productivity, and at the same time in taking care of its employees' health.

Vibration

Vibration is the oscillatory motion of various bodies. All bodies with mass elements and elasticity are capable of vibration; hence, most machines and structures including the human body experience vibration to some degree. Two different categories of vibration are distinguished in literature. Free vibration takes place when the system oscillates due to the action of internal forces only. Forced vibration is caused by the action of external forces. If the frequency of excitation coincides with the natural frequency of the system, resonance occurs. The result is large oscillations within the structure of creating potentially harmful stress. For example, the potential effect of resonance is the shattering of a crystal glass, when opera singer sings at the natural frequency of the crystal. Energy dissipation leads to friction and other resistances, damping occurs in all structures.

The Effects of Vibration on Human Body

Vibrations affect human body in many different ways. The response to a vibration exposure is primarily dependent on the frequency, amplitude, and duration of exposure. Other factors may include the direction of vibration input, location and mass of different body segments, level of fatigue and the presence of external support. The human response to vibration can be both mechanical and psychological. Mechanical damage to human tissues can also occur as they are caused by resonance within various organ systems. Psychological stress reactions also occur from vibrations; however, they are not necessarily frequently related. From point of view of exposure, the low frequency range of vibration is the most interesting. Exposure to vertical vibrations in the 5-10 Hz range generally causes resonance in the thoracic-abdominal system, at 20-30 Hz in the head-neck-shoulder system, and at 60-90 Hz in the eyeball. When vibrations are attenuated in the body, its energy is absorbed by the tissues and organs. The muscles are also important in this respect.

Vibration leads to both voluntary and involuntary contractions of muscles, and can cause local muscle fatigue, particularly when the vibration is at the resonant-frequency level. Furthermore, it may cause reflex contractions, which will reduce motor performance capabilities. The amount of mechanical energy transmitted due to vibrations depends on the body position and muscle contractions. In a standing subject, the first resonance occurs at the hip, shoulder, and head at about 5Hz. With sitting subjects, resonance occurs at the shoulders and to some degree at the head at 5 Hz. Furthermore, a significant resonance oceans from shoulder to head occurs at about 30 Hz.

Based on psychological studies, observations indicate that the general state of consciousness is influenced by vibrations. Low frequency vibrations 1-2 Hz with moderate intensities induce sleep. Unspecific psychological stress reactions have also been noted (Guignard & Von Gierke), as well as degraded visual and motor effects on functional performance. Some symptoms of vibration exposure at low frequencies are given in Table 1, along with the frequency ranges at which the symptoms are most predominant.

Table 5.2 Symptoms caused by Whole-Body Vibration and the Frequency Range at which They Usually Occur (Adapted from Rasmussen, 1982).

Symptoms	Frequency (Hz)
General feeling of discomfort	4-9
Head symptoms	13-20
Lower Jaw symptoms	6-8
Influence on speech	13-20
"Lump in throat"	12-16
Chest Pain	5-7
Abdominal pain	4-10
Fregment Urge to urinate	10-18
Increased muscle tone	13-20
Influence on breathing movements	4-8
Muscle contractions	4-9

Summary

This chapter deals with the introduction, history of development and objectives of Ergonomics It also discusses the various types of man-machine systems, their design characteristics and classifications. Apart from this, it also explains various measures of physiological functions, Applied Anthropometry, design of seat and workplace, visual display design, and environment risk factors.

❑❑❑

Index

❑❑❑